AF728945

Electromagnetic Processing of Materials

FLUID MECHANICS AND ITS APPLICATIONS
Volume 99

Series Editor: R. MOREAU
MADYLAM
Ecole Nationale Supérieure d'Hydraulique de Grenoble
Boîte Postale 95
38402 Saint Martin d'Hères Cedex, France

Aims and Scope of the Series

The purpose of this series is to focus on subjects in which fluid mechanics plays a fundamental role.

As well as the more traditional applications of aeronautics, hydraulics, heat and mass transfer etc., books will be published dealing with topics which are currently in a state of rapid development, such as turbulence, suspensions and multiphase fluids, super and hypersonic flows and numerical modeling techniques.

It is a widely held view that it is the interdisciplinary subjects that will receive intense scientific attention, bringing them to the forefront of technological advancement. Fluids have the ability to transport matter and its properties as well as to transmit force, therefore fluid mechanics is a subject that is particularly open to cross fertilization with other sciences and disciplines of engineering. The subject of fluid mechanics will be highly relevant in domains such as chemical, metallurgical, biological and ecological engineering. This series is particularly open to such new multidisciplinary domains.

The median level of presentation is the first year graduate student. Some texts are monographs defining the current state of a field; others are accessible to final year undergraduates; but essentially the emphasis is on readability and clarity.

For further volumes:
http://www.springer.com/series/5980

Shigeo Asai

Electromagnetic Processing of Materials

Materials Processing by Using Electric and Magnetic Functions

Springer

Shigeo Asai
Japan Science and Technology Agency
Ahara-cho 23-1
457-0063 Nagoya
Minami-ku
Japan
asai@numse.nagoya-u.ac.jp

This work is partly based on an updated translation from the original Japanese work entitled: "Zairyou Denzi Purossesingu", by Shigeo Asai, 2000, Uchida Rokakuho, Japan.

ISSN 0926-5112
ISBN 978-94-007-2644-4 e-ISBN 978-94-007-2645-1
DOI 10.1007/978-94-007-2645-1
Springer Dordrecht Heidelberg London New York

Library of Congress Control Number: 2011945107

© Springer Science+Business Media B.V. 2012
No part of this work may be reproduced, stored in a retrieval system, or transmitted in any form or by any means, electronic, mechanical, photocopying, microfilming, recording or otherwise, without written permission from the Publisher, with the exception of any material supplied specifically for the purpose of being entered and executed on a computer system, for exclusive use by the purchaser of the work.

Printed on acid-free paper

Springer is part of Springer Science+Business Media (www.springer.com)

Preface

Most people who choose to read this book are likely to begin with the question "What is ***Electromagnetic Processing of Materials?***" This question is a reasonable one since ***Electromagnetic Processing of Materials,*** so called ***EPM***, will find its future development in concert with ***Materials Science and Engineering***, a conjoining of areas of research that has not yet been fully consummated.

In the metals industry, utilization of the *Lorentz* force started in very early times; for example, machines to effect electromagnetic levitation and electromagnetic mixing were invented in 1923 and 1932, respectively, but with little understanding of the science behind them. To bridge the gap between technology and science, the term ***Magnetohydrodynamics***, which had been established by *Alfvén* in 1942, was first introduced at the IUTAM (International Union of Theoretical and Applied Mechanics) symposium entitled "Metallurgical Applications of Magnetohydrodynamics", held in 1982 in Cambridge, England. Thus, we can say that ***Electromagnetic Processing of Materials*** started from this symposium where the two channels of ***Metallurgy*** and ***Magnetohydrodynamics*** were combined. The term ***Electromagnetic Processing of Materials*** first appeared at The Sixth International Iron and Steel Congress held in 1990 in Nagoya, Japan and then the first Symposium of ***EPM*** was held in 1994, also in Nagoya. That is, ***EPM*** started in the 1980s and grew in the 1990s as an intimate collaboration between France and Japan. The first fruits of EPM progress were found in utilization of the *Lorentz* force, mainly for development of steelmaking processes, especially of continuous casting processes. By the 2000s the activity of ***EPM*** had spread into the utilization of magnetization force, enabled by the fortuitous fact that a high magnetic field, up to around 15 Tesla, became easily available through the development of super-conducting technologies. Briefly speaking, ***EPM*** is now located in one branch of ***Materials Science and Engineering*** and aims to create new materials and/or to design efficient processes by making use of various functions which appear when electric and magnetic fields are applied to known materials.

This book contains not only an English translation of my book on Electromagnetic Processing of Materials first published in 2000 by Uchida Rokakuho, Japan, the content of which is based on the lectures given for advanced undergraduate and graduate students in the Department of Materials Science and Engineering, Nagoya University over two decades, but also the introduction of transport phenomena. Furthermore, tables and appendix are added for helping the design of electromagnetic processes, and better understanding of the contents of the book. In Chap. 1, the birth, characteristics and prospects of ***EPM*** are described to give readers a brief summary of what ***EPM*** is. In Chap. 2, the concept of ***Transport Phenomena,*** which is indispensable in materials processing, is given a concise introduction in order to establish that the transport phenomena of mass, heat and momentum can all be described by the same diffusion type equation. In Chap. 3, I describe the essentials of ***Magnetohydrodynamics*** that are required for understanding ***EPM*** and I show that the governing equation of a magnetic field can also be expressed by the same diffusion type equation as introduced in Chap. 2. That is, it is understood that the concept of transport phenomena holds not only in relation to heat, mass and momentum, but also in magnetic fields. Though many mathematical equations appear in this chapter, for the sake of developing a stricter analysis, I recommend that the beginner in EPM read until Chap. 5 by skipping derivation of equations. After that, readers should again read the skipped Chapters. In Chap. 4, the various functions that appear when applying electric and magnetic fields to materials are enumerated and their utilization in materials processing is explored. Recently, due to advances in superconducting magnets, a high magnetic field has become readily available and is being applied in various fields of science. In this trend many interesting phenomena relating to high magnetic fields have been found and a new academic area called ***Magneto-Science*** has begun to open a gate to further research. In order to connect the seeds sprouting from the new academic field with the needs of ***Materials Science and Engineering***, a new branch of ***EPM*** is growing. Thus, in Chap. 5, materials science and engineering utilizing a high magnetic field is introduced and the functions that appear when applying a high magnetic field to materials processing are explained.

EPM has only recently been recognized as a cutting edge technology. If publication of this textbook leads to opening a new academic field in ***Materials Science and Engineering***, and becomes universally known as ***Electromagnetic Processing of Materials,*** it will deepen my great pleasure as one of the persons who have participated in this field from the beginning. Especially, I would like to dedicate this book to the late Dr. Marcel Garnier of the EPM-MADYRAM in Grenoble, France, who first initiated and developed this field together with me.

I am indebted to many people for help of one kind or another in the making of this book. I am especially grateful to the late Professor Iwao Muchi at Nagoya University and the late Professor Julian Szekely at MIT, who taught me the basis of **Metallurgical Reaction Engineering** and **Magnetohydrodynamics**, respectively. Those academic backgrounds prompted me to write this book. Professor Emeritus Tsuyoshi Masumoto of Tohoku University, Japan always has encouraged us in developing this field by nominating the EPM subject to a place in the science and technology program in The Ministry of Education, Culture, Sports, Science and Technology, Japan. Professor Emeritus René Moreau of SIMAP/EPM in Grenoble, not only gave useful academic advice regarding the problems of Magnetohydrodynamics, but also kindly arranged a contact with Springer to publish this book. The Professor Yves Fautrelle of SIMAP/EPM in Grenoble and Professor Shoji Taniguchi of Tohoku University have devoted a lot of effort for developing EPM with me for a long time. Above all, the author would like to thank Professor Ken-ichi Miyazawa of Nagoya University who painstakingly read each draft, querying every ambiguity and exposing the many inconsistencies in the original manuscript. Careful reviews of the final manuscripts contributed crucial descriptions of the academic viewpoints of Profs. Noboru Yoshikawa and Kazuyuki Ueno of Tohoko University, Kazuhiko Iwai of Nagoya University and Hideyuki Yasuda of Osaka University. In this sense I am not the only author of this book, but share that role with all of my colleagues in world EPM community.

Despite this help, errors and obscurities undoubtedly remain. I would be most grateful to anyone who brings them to my attention at one of the addresses below.

Postal address: Yakushiyama 112, Narumi-cho,
Midori-ku, Nagoya, 458-0846, Japan
e-mail address: asai@numse.nagoya-u.ac.jp
and shigeo-asai@tg.commufa.jp

Japan Shigeo Asai

Contents

Chapter 1
Development of Electromagnetic Processing of Materials

The history, scientific basis, industrial background and academic view of Electromagnetic Processing of Materials are outlined here.

1 The Birth of Electromagnetic Processing of Materials

Science scholars say that the roots of "Metallurgy" can be found in the book titled "De re metallica" written by Georg Agricola, 1556. On the other hand, the foundations of Fluid Dynamics were not laid until the eighteenth and nineteenth centuries with the fundamental contributions of Navier and Stokes, and the field of Electric and Magnetic Science was established by Maxwell as late as 1878. In the metallics industry, electric energy has for many years been used as heat in melting, refining and solidifying processes. It is surprising to realize that technologies using electric energy were developed in a rather early era; for instance, electromagnetic levitation, the concept of a cold crucible and electromagnetic stirring were all invented in 1923, 1932, and 1931, respectively. These technologies were developed well ahead of the back-ground of scientific understanding on which their efficacy is based. It is now known that imposing an electric current on a molten metal induces a magnetic field, and an electromagnetic force is as a result generated by coupling of the induced magnetic field and the imposed electric current. This electromagnetic force causes motion of the molten metal. The same thing occurs when imposing an alternating magnetic field on a molten metal, that is, the molten metal is driven to flow. The appearance that the molten metal is driven by imposing an electric current and/or a magnetic field are well-known phenomena that indicate that the motion of an electrically conductive fluid and the electric and magnetic fields act upon each other. The scientific theory that encompasses the phenomena coupling Fluid Dynamics and Electric and Magnetic Science is called Magnetohydrodynamics (MHD), a theory that has been developed in such fields of science and engineering

S. Asai, *Electromagnetic Processing of Materials*, Fluid Mechanics and Its Applications 99, DOI 10.1007/978-94-007-2645-1_1,
© Springer Science+Business Media B.V. 2012

as plasma physics, geophysics, astronomy, nuclear fusion and nuclear power. We briefly review the history of MHD [1] in the following.

At the beginning of the nineteenth century, Faraday thought that motions of the sea might account for the observed perturbations of earth's magnetic field. Over the period from the late nineteenth century to the beginning of the twentieth century, it was suggested that the magnetic fields of the sun and the other heavenly bodies might be due to *dynamo action,* whereby the electrically conductive materials of stars acted as the armature and stator of a self-exciting dynamo. Finally in 1942 the engineer-astrophysicist Alfvén (Nobelist in 1970) established MHD. Indeed the term "magnetohydrodynamics" itself was coined by him about this time.

The research institute called MADYLAM was established in Grenoble, France in 1978 to apply MHD to materials processing. To provide a scientific understanding for the technologies developed in the metal industry, MHD was first introduced at the IUTAM conference [2] titled on "The Application of Magnetohydrodynamics to Metallurgy", held at Cambridge, England, 1982. This symposium first developed the theme of metallurgical applications of MHD. In Japan, encouraged by the symposium, the Iron and Steel Institute of Japan (ISIJ) first decided to study an area of MHD that they identified as "Electromagnetic Metallurgy". Later, in 1985, the ISIJ formed the Committee of Electromagnetic Metallurgy, which was replaced in 1989 by the Committee of Electromagnetic Processing of Materials, and established a session under that title in the 6th International Iron and Steel Congress held in Nagoya, Japan, 1990. The first Symposium on EPM was held in Nagoya, 1994. That is, EPM was formulated in the 1980s and grew in the 1990s with strong support of the ISIJ and a tight collaboration between France and Japan. At first EPM grew in utilization of the Lorentz force mainly relating to the development of steel-making processes, especially in continuous casting of steels. Recently the research activity on EPM has found another application of the magnetization force, under the comfortable circumstance where a high magnetic field up to around 15T has been easily obtainable thanks to the development of super-conducting technologies.

The roots and progress of EPM are summarized in Table 1.1.

2 The Comparison of Various Energies

The amount of various energies used in materials processing is calculated by the use of characteristic values and listed in Table 1.2. From the table, we can understand that the field of motion is influenced by a magnetic field since both magnetic and motion energies are almost the same value of $\sim 10^5$ J/m^3. Also it can be seen that this magnetic energy is too small in comparison with the thermal energy of $\sim 10^{10}$ J/m^3. Thus, the physical values determined by the thermal energy cannot be changed by imposition of a magnetic field. On the other hand, the magnetization energy

Table 1.1 The roots and progress of EPM

Metallurgy (1556)
De re metallica by Agricola
↓
Levitation Melting (1923)
↓
Electromagnetic Stirring (1932)

Hydrodynamics (18-19th Cen.)
↓
Electromagnetism (1873)
↓ **by Maxwell**
MHD Phenomena (1889, 1919)
↓
Magnetohydrodynamics (1942)
by Alfven

↘ **Electromagnetic Processing of Materials** ↙

EPM activities

MADYLAM	···	**France (1978)**
IUTAM Symposium	···	**England (1982)**
Committee of EPM in ISIJ	···	**Japan (1985)**
The 1st Symp. of EPM	···	**Japan (1994)**
The 2nd Symp. of EPM	···	**France (1997)**
The 3rd Symp. of EPM	···	**Japan (2000)**
The 4th Symp. of EPM	···	**France (2003)**
The 5th Symp. of EPM	···	**Japan (2006)**
The 6th Symp. of EPM	···	**Germany (2009)**

Table 1.2 A comparison with various energies

Magnetic energy density in vacuum

$$W=\frac{1}{2}\boldsymbol{H}\cdot\boldsymbol{B}=\frac{1}{2\mu_m}\,|\,\boldsymbol{B}\,|^2=4\times10^5\ \mathrm{J/m^3}$$

Thermal energy density (molten steel)

$$W=c_p\rho T=10^{10}\ \mathrm{J/m^3}$$

Electric energy density (electric strength in air: E = 1 kV / mm)

$$W=\frac{1}{2}\boldsymbol{E}\cdot\boldsymbol{D}=\frac{1}{2}\varepsilon\,|\,\boldsymbol{E}\,|^2=4.4\ \mathrm{J/m^3}$$

Motion energy density ($V = 5\ \mathrm{m/s}$, $\rho = 7.8\times10^3\ \mathrm{kg/m^3}$)

$$W=\frac{1}{2}\rho V^2=10^5\ \mathrm{J/m^3}$$

Magnetization energy density (non magnetic material, $B = 10\mathrm{T}$)

$$W=\frac{1}{2\mu_m}\chi\,|\,\boldsymbol{B}\,|^2=\frac{1}{2\times4\pi\times10^{-7}}(10^{-3}\sim10^{-4})\times10^2$$
$$=(4\sim0.4)\times10^4\ \mathrm{J/m^3}$$

is increased up to $\sim10^4\ \mathrm{J/m^3}$ by imposing a high magnetic field of 10T to non-magnetic materials, so we can see that the high magnetic field will be a useful tool in materials processing. This matter will be discussed more precisely in Chap. 5.

3 Industrial Background and Engineering Aspects

3.1 *Increase of Electric Energy Consumption*

A strong demand for high-grade materials is always accompanied by an increase in the consumption of electric energy. Especially, as steelmaking continues to progress, the number of treatments available to refine steels is increasing due to the demand for ever cleaner steels. Therefore, the time needed for refining tends to be prolonged and addition of thermal energy to a molten steel has been an important technical subject; i.e., how to compensate for heat loss during the prolonged refining time. This thermal energy is mainly supplied by electric energy.

3.2 *Development of Related Technologies*

The technology connected with electricity and magnetism has seen dramatic developments. For instance, the recently produced high-energy permanent magnet (e.g., Fe-Nd-B) has easily given us a strong magnetic field with a cheap cost. Also, superconducting magnets have been introduced in various research fields in universities and applied even on the test plant of a continuous casting of steels in Japan. Moreover, an electrode made of ZrB_2 (melting temperature: 3,000 C and electrical resistivity: 10^{-5} $\Omega{\cdot}$m) has been developed. This has opened the gate to impose electric current directly on molten metals with high melting temperature such as molten steel. Until now the electric current induced by an alternating magnetic field has been mainly adopted in the metal industry (e.g. linear type electromagnetic stirring, electromagnetic pump, etc.) instead of directly imposing electric current on molten metals. From now on, by directly imposing electric current we can expect alternative methods for transport and mixing of molten metals with better efficiency of electric power.

3.3 *High Density, Cleanliness and Controllability of Electric Energy*

Making use of an electric current and a magnetic field is the most convenient method to impose energy with high density on materials. The application of a magnetic field is completely free from contamination and the imposition of electric current is extremely clean, except for contamination from electrodes. Furthermore, the controlling technique with respect to electric current and magnetic field has been much advanced in recent times.

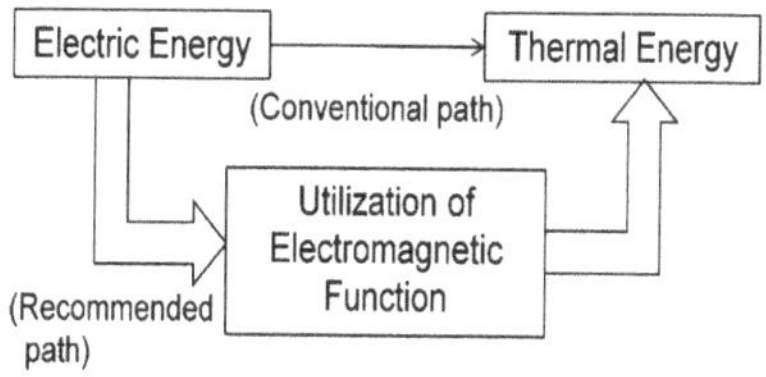

Fig. 1.1 The recommended energy path and the conventional path

Table 1.3 A comparison of physical characteristics between plasma and molten metal

Plasma	Compressible
	Anisotropic (steep temp. gradient for the plasma used in materials processing)
Molten metals	Incompressible
	Isotropic (uniform temp. field due to small Prandtle number)

3.4 *Effective Use of Electric Energy*

In the metals refining processes, electric energy has been mainly used as a heat source. The heating rate required as heat addition in steelmaking processes is about $1 \sim 6$ C/min. On the other hand, the values of ordinary mixing power employed in the steel-making processes are $1 \sim 500$ W/ton, which corresponds to a heating rate of about 1 C/a week $\sim$1 C/30 min.. Thus, it can be seen that electric power for heat addition is $30 \sim 60{,}000$ times larger than the required mixing power. Therefore, as shown in Fig.1.1, part of the electric energy should be converted into kinetic energy in a way that will add some functions of electromagnetism such as mixing, splashing, driving, etc., the details of which will be discussed in Chap. 4. The energy used for functions of electromagnetism is not wasted but is finally converted into thermal energy through viscous dissipation.

3.5 *Application of Magnetohydrodynamics to Materials Processing*

Magnetohydrodynamics, which was developed in a field treating the dynamics of electrically conductive fluid such as plasma, of course includes the dynamics of molten metals. The comparison between plasma and molten metals with fluidic characteristics is shown in Table 1.3. The vast knowledge stored in magnetohydrodynamics is now applicable to solve problems in processing of molten metals. The advantage of this application is the simplifying procedure from a compressible and anisotropic fluid to an incompressible and isotropic one.

4 Tree of Electromagnetic Processing of Materials

Figure 1.2 reveals an overview of the electromagnetic processing of materials as a tree. The roots indicate the academic background supporting this engineering field as follows. The branches predict functions of electromagnetism and the leaves in each branch show processes and technologies related to the corresponding function as described in later chapters. The Electromagnetic Processing of Materials (EPM) is based on Magnetic Science, Process Metallurgy and Magnetohydrodynamics (MHD), where the functions of electromagnetism are utilized for processing of various materials including the electrically conductive substances.

- Electromagnetic processing of materials
  - Magnetic science
  - Process metallurgy
    - Transport phenomena
    - Thermodynamics
  - Magnetohydrodynamics
    - Electromagnetism
    - Fluid mechanics

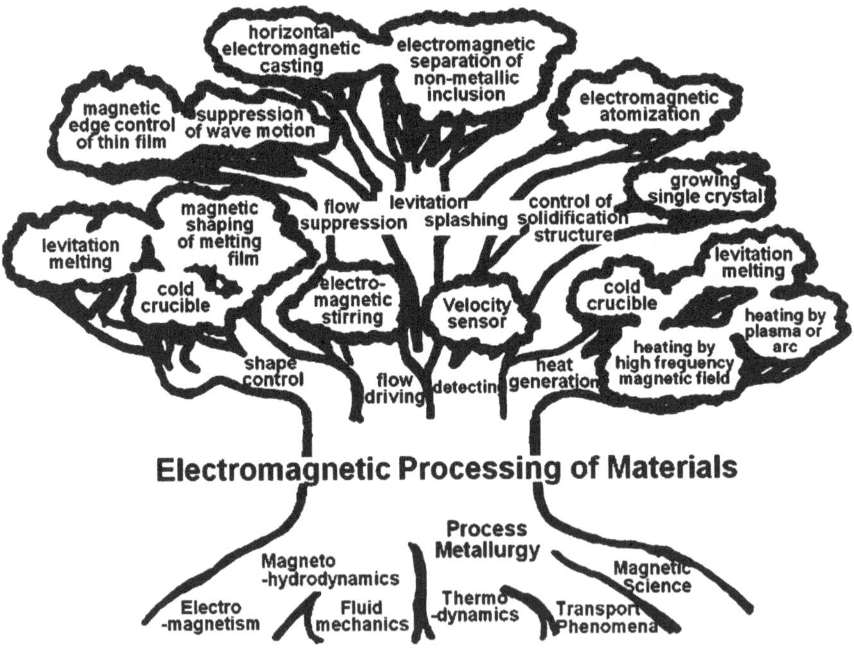

Fig. 1.2 A tree of electromagnetic processing of materials

References

1. J.A. Shercliff, *A Textbook of Magnetohydrodynamics* (Pergamon Press, Oxford, 1965)
2. H.K. Moffatt, M.R.E. Proctor, Metallurgical application of magnetohydrodynamics, in *Proceedings of a Symposium of the IUTAM*, The Metal Society, London, 1984

Chapter 2
Transport Phenomena in Materials Processing

Before studying Electromagnetic Processing of Materials, let us see how the transport phenomena of mass, heat and momentum can be formulated in mathematical expressions. Later we will learn that a magnetic field can be described as a transport phenomenon where the governing equation of the magnetic field is the same type equation as those appearing in the transport phenomena of mass, heat and momentum, i.e., the partial differential equation of parabolic type.

1 Formulation

1.1 Flux

A flux is defined as a physical quantity flowing through a unit cross-sectional area perpendicular to its flowing direction per unit time period. The physical quantities treated in Transport Phenomena are mass, energy and momentum. The signs and units of each physical quantity are given as follows:

Mass flux	$\boldsymbol{J}\ \{kg/(m^2\cdot s)\}$	Vector
Heat (Energy) flux	$\boldsymbol{q}\ \{J/(m^2\cdot s)\}$	Vector
Momentum flux	$\boldsymbol{\tau}\ \{(kg\cdot m/s)/(m^2\cdot s)\}$	Second-order tensor

The term "flux" means "flow per unit area". The mass flux $\boldsymbol{J}$ is the mass amount (kg) flowing through unit area (m^2) per unit time period (s) so that the dimension expressed in the MKS system is $\{kg/(m^2\cdot s)\}$. In the same way, the dimensions of the heat (energy) flux $\boldsymbol{q}$ and the momentum flux $\boldsymbol{\tau}$ are $\{J/(m^2\cdot s)\}$ and $\{(kg\cdot m/s)/(m^2\cdot s)\}$ in the MKS system, respectively. As the mass and heat fluxes express both transport direction and amount, those are vector quantities. On the other hand, the momentum flux is interpreted as itself having a vector component in the transported direction so that it is a second-order tensor quantity.

S. Asai, *Electromagnetic Processing of Materials*, Fluid Mechanics and Its Applications 99, DOI 10.1007/978-94-007-2645-1_2,
© Springer Science+Business Media B.V. 2012

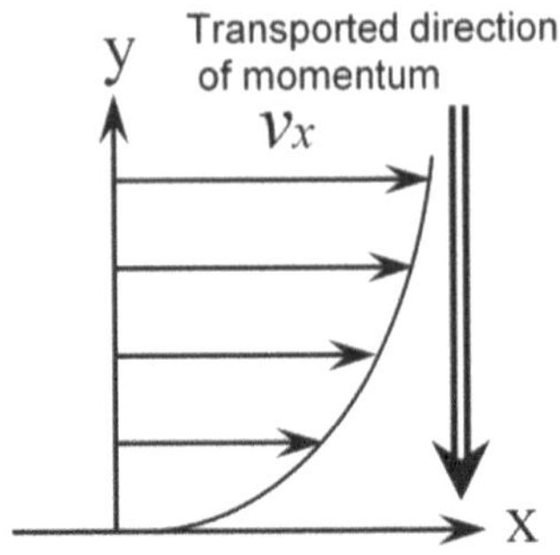

Fig. 2.1 Momentum transport

The diffusion transports of mass, heat, and momentum are formulated by Eqs. 2.1, 2.2, and 2.3 which are known as *Fick's first law, Fourier's law of heat conduction* and *Newton's law of viscosity*, respectively.

$$J_x = -D(\partial c/\partial x), \tag{2.1}$$

where D is the diffusion coefficient having the unit of (m^2/s) and c the concentration of a chemical species. Equation 2.1 indicates that the flux in the x-direction J_x is in proportion to the gradient in the x-direction of the concentration c and the diffusion coefficient D. The (−) sign appearing in the right hand side of Eq. 2.1 implies that a substance is transported to the negative direction of the concentration gradient, i.e., the substance moves from a higher concentration point to a lower one.

$$q_x = -\lambda(\partial T/\partial x), \tag{2.2}$$

where λ is the thermal conductivity having the unit of $\{J/(m{\cdot}s{\cdot}K)\}$ and T the temperature. Like the mass transport, the (−) sign also indicates that the heat moves from a higher temperature point to a lower one,

$$\tau_{yx} = -\mu(\partial v_x/\partial y), \tag{2.3}$$

where μ is the viscosity coefficient with the unit of $\{kg/(m{\cdot}s)\}$. This equation states that the shearing force per unit area is in proportion to the negative velocity gradient and the proportional coefficient (viscosity coefficient) μ, i.e., the momentum τ also is transported from a higher to a lower velocity point, as were the previous physical quantities. In the neighborhood of a fixed solid wall at $y=0$ the fluid with the velocity v_x exerts a certain amount of x-momentum and transports the momentum through the adjacent layer of fluid to the fixed solid wall, as shown in Fig. 2.1. Namely the x-momentum is being transported through the fluid in the negative y-direction. Therefore τ_{yx} may also be interpreted as the flux of the x-momentum in the y-direction.

1.2 Conservation Law

The conservation law holds for the physical quantities of mass, heat and momentum, which we discuss in this chapter. That is;

Mass :	Law of conservation of mass
Heat (Energy) :	Law of conservation of energy
Momentum :	Law of conservation of momentum

These laws are formulated in Eq. 2.4, which is developed by writing the physical quantity α in a volume element $\Delta x \Delta y \Delta z$ fixed in space, through which the α is transported.

$$\begin{aligned} \text{A [rate of accumulation of } \alpha] &= \text{B [rate of } \alpha \text{ flowing into } \Delta x \Delta y \Delta z] \\ &\quad - \text{C [rate of } \alpha \text{ flowing out of } \Delta x \Delta y \Delta z] \\ &\quad + \text{D [rate of increase(decrease) of } \alpha] \end{aligned} \tag{2.4}$$

By the way, this conservation law also holds for the dynamics of money. In the case of a bank account, A = change of accounted amount in a period, B = deposited amount in the period, C = withdrawn amount in the period, and D = interest amount in the period. This is reasonable since money is one of manifestation of the concept of energy.

1.3 Mathematical Formulation and Solution

Using Eqs. 2.1, 2.2, and 2.3, which express the diffusion transports of mass, heat and momentum, and the conservation law of Eq. 2.4, one can derive the governing equations for the transports of mass, heat and momentum. The space distribution and time change of each physical property is obtained by solving the governing equations under reasonable boundary and initial conditions. The solution(s) can be categorized as analytical or numerical. In general, it is not easy to get an analytical solution without using a high-level mathematical technique even in a less complicated system. We will later introduce a dimensional analysis that will be useful in such cases. A numerical solution can also be a useful method, but the over-all grasp of the physical concept is not easy to obtain when using existing package programs. The procedure for finding solutions is summarized as follows:

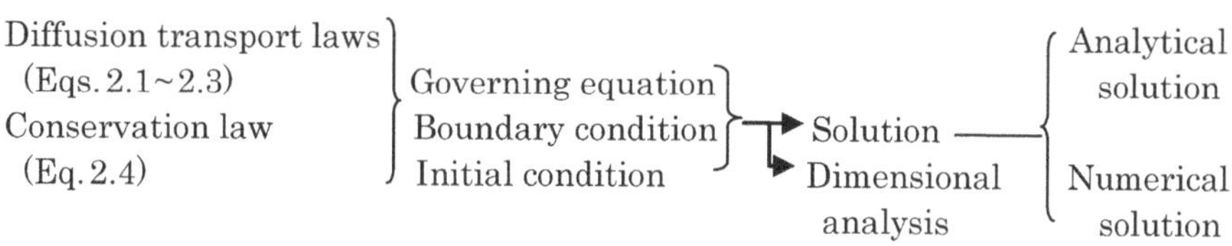

2 Derivation of Governing Equations

Before deriving the governing equations, let us look at some coordinate systems. In this text we only use the familiar rectangular (Cartesian), circular cylindrical, and spherical coordinates systems. By convention, a right-handed coordinate system is always used whereby one curls the fingers of his or her right hand in the direction from x to y so that the forefinger is in the x-direction and the middle finger is in the y-direction. The thumb then points in the z-direction, as shown in part **a** in Fig. 2.2. This convention is necessary to remove directional ambiguities in theorems to be derived later. In the other coordinate systems, the coordinates indicated by the respective fingers of the right hand are shown in parts **b** and **c** in Fig. 2.2. Hereafter, let us derive the equation of continuity, the equation of continuity for species, the equation of change for temperature, and the equation of motion in rectangular coordinates by applying Eq. 2.4 to a small volume element $\Delta x \Delta y \Delta z$. Furthermore, in order for the procedure to be understood more easily, the equations will be

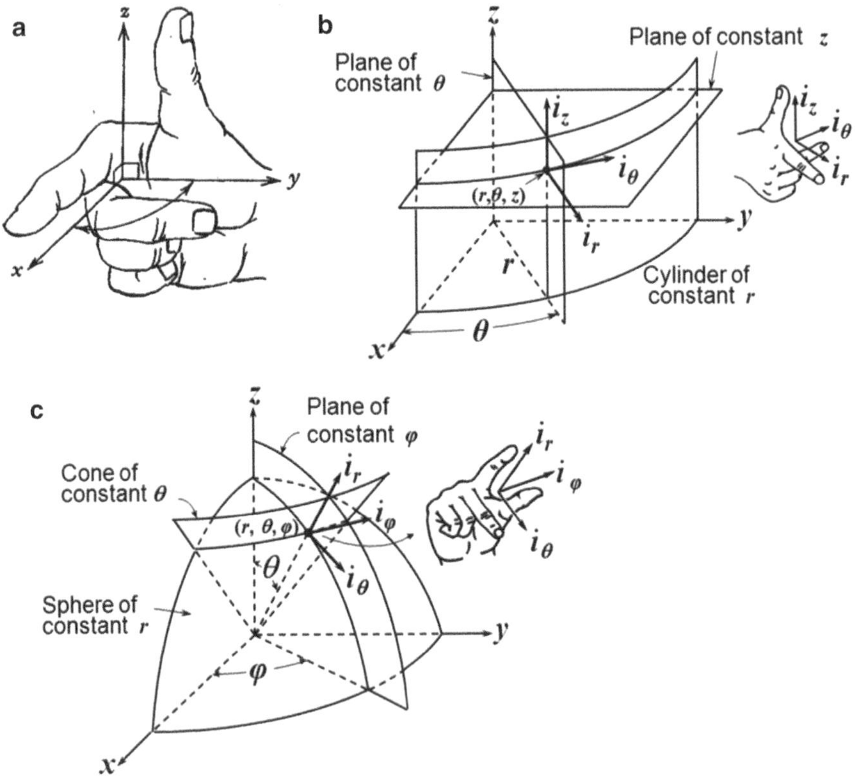

Fig. 2.2 Coordinate Systems. **a** Rectangular (Cartesian) coordinates, **b** Circular cylindrical coordinates and **c** Spherical coordinates [1]

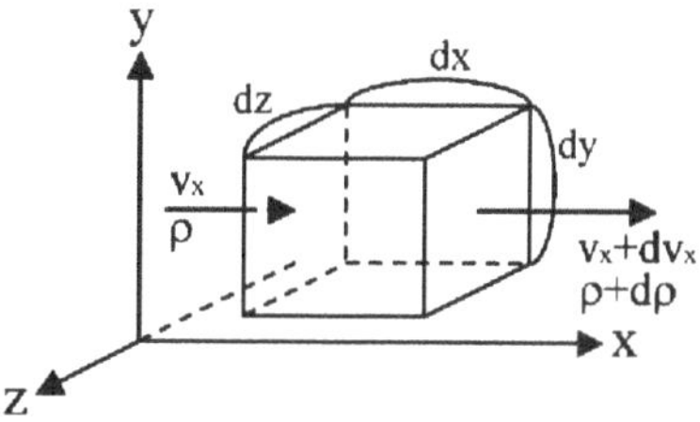

Fig. 2.3 Mass balance of a fluid in one dimension

derived under the conditions of one-directional flow and transport of each physical property. The equations will then be extended to the general governing equations, which are not limited to the conditions of one-directional flow and transport and the rectangular coordinates employed here.

2.1 *The Equation of Continuity*

Let us consider the system where only the x-component flow v_x exists, and take the mass balance in a small element $\Delta x\Delta y\Delta z$ (see Fig. 2.3. The terms of A, B, C, D in Eq. 2.4 are obtained as follows:

$$\left.\begin{aligned} &\mathrm{A} = \partial(\rho\Delta x\Delta y\Delta z)/\partial t \\ &\mathrm{B} = \rho v_x(\Delta y\Delta z) \\ &\mathrm{C} = (\rho + \Delta\rho)(v_x + \Delta v_x)(\Delta y\Delta z) \\ &\mathrm{D} = 0 \end{aligned}\right\}, \tag{2.5}$$

where Δv_x is the variation of v_x in the x-direction and expressed by $(\partial v_x/\partial x)\Delta x$, and $\Delta\rho$ is the variation of ρ in the x-direction and expressed by $(\partial\rho/\partial x)\Delta x$ Substituting those terms into Eq. 2.4 yields Eq. 2.6.

$$(\partial\rho/\partial t)\Delta x\Delta y\Delta x = \rho v_x(\Delta y\Delta z) - \{\rho + (\partial\rho/\partial x)\Delta x\}\{v_x + (\partial v_x/\partial x)\Delta x\}(\Delta y\Delta z) \tag{2.6}$$

When both sides of Eq. 2.6 are divided by $\Delta x\Delta y\Delta z$ and the limit is taken as Δx approaches zero, we get the equation of continuity in its one-dimensional form as Eq. 2.7.

$$(\partial\rho/\partial t) + \partial(\rho v_x)/\partial x = 0 \tag{2.7}$$

A very important special form of the equation of continuity is Eq. 2.8 which holds for a fluid of constant density (incompressible fluid).

$$\partial v_x/\partial x = 0 \tag{2.8}$$

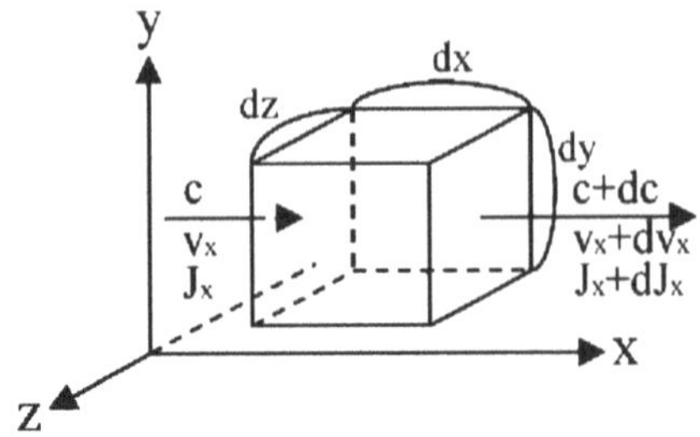

Fig. 2.4 Mass balance of a species in a fluid in one dimension

If we take account of existence of not only v_x but also v_y and v_z in the derivation of Eq. 2.7, the equation of continuity in the three dimensional form is obtained as Eq. 2.9.

$$(\partial\rho/\partial t) + \partial(\rho v_x)/\partial x + \partial(\rho v_y)/\partial y + \partial(\rho v_z)/\partial z = 0 \tag{2.9}$$

In the case of an incompressible fluid, Eq. 2.9 reduces to Eq. 2.10.

$$\partial v_x/\partial x + \partial v_y/\partial y + \partial v_z/\partial z = 0, \qquad \nabla \cdot \boldsymbol{v} = 0, \tag{2.10}$$

where ∇ means the vector differential operator, called "nabla".

Next, we apply the law of conservation of mass for species. The system we consider here is a volume element $\Delta x \Delta y \Delta z$ fixed in space, through which the fluid mixture flows in the x-direction (see Fig. 2.4). The terms of A, B, C and D in Eq. 2.4 are given as follows:

$$\left.\begin{aligned} &\mathrm{A} = \partial(c\Delta x\Delta y\Delta z)/\partial t \\ &\mathrm{B} = (v_x c + J_x)(\Delta y\Delta z) \\ &\mathrm{C} = \{(v_x + \Delta v_x)(c + \Delta c) + (J_x + \Delta J_x)\}(\Delta y\Delta z) \\ &\mathrm{D} = 0 \end{aligned}\right\}, \tag{2.11}$$

where Δv_x, Δc, and ΔJ_x are the variation of v_x, c and J_x in the x-direction, respectively, and expressed as follows:

$$\Delta v_x = (\partial v_x/\partial x)\Delta x, \;\; \Delta c = (\partial c/\partial x)\Delta x,$$

$$\text{and } \Delta J_x = [\partial\{-D(\partial c/\partial x)\}/\partial x]\ \Delta x = -D(\partial^2 c/\partial x^2)\Delta x$$

By substituting A, B, C and D in Eq. 2.11 into Eq. 2.4, Eq. 2.12 is obtained.

$$\underbrace{(\partial c/\partial t)\Delta x\Delta y\Delta z}_{\mathrm{A}} = \underbrace{\{v_x c - D(\partial c/\partial x)\}(\Delta y\Delta z)}_{\mathrm{B}}$$
$$-\underbrace{[\{v_x + (\partial v_x/\partial x)\Delta x\}\{c + (\partial c/\partial x)\Delta x\} + \{-D(\partial c/\partial x) - D(\partial^2 c/\partial x^2)\Delta x\}](\Delta y\Delta z)}_{\mathrm{C}} \tag{2.12}$$

When this equation is divided by $\Delta x \Delta y \Delta z$ and the limit is taken as Δx approaches zero, we get *the equation of continuity for one species* in its one-dimensional form as Eq. 2.13.

$$\partial c/\partial t + v_x(\partial c/\partial x) = D(\partial^2 c/\partial x^2) \tag{2.13}$$

By taking account of mass transport in the *x*-, *y*- and *z*-directions, Eq. 2.14 is obtained.

$$\begin{aligned} \partial c/\partial t + v_x(\partial c/\partial x) + v_y(\partial c/\partial y) + v_z(\partial c/\partial z) \\ = D(\partial^2 c/\partial x^2 + \partial^2 c/\partial y^2 + \partial^2 c/\partial z^2) \\ \text{or } Dc/Dt \equiv \partial c/\partial t + \boldsymbol{v}\cdot\nabla c = D\nabla^2 c, \end{aligned} \tag{2.14}$$

where ∇^2 is called a Laplacian operator both for scalar and vector fields, expressed by $\nabla^2 = \partial^2/\partial x^2 + \partial^2/\partial y^2 + \partial^2/\partial z^2$ in Cartesian coordinates. Equation 2.14 is *the equation of continuity for one species* and called the equation of mass balance in its three-dimensional form.

Problem II-1. Following the derivation of Eq. 2.13, derive *the equation of continuity for one species* in its one-dimensional form in the *y*-direction. Assume only the flow in the y-direction v_y exists.

2.2 Energy Equation for a Fluid with Constant Density

In principle, the conservation of energy should include internal, kinetic, electromagnetic, chemical forms of energy, etc. Under several restricted conditions including Fourier's law with constant λ and Newton's law, the equation of change for temperature can be employed as the most basic and simplest form of the energy equation [2]. This type of energy equation can be derived by taking the same procedure in the derivation of Eq. 2.14, where the concentration *c* and J_x are replaced by the heat capacity per unit volume $\rho c_p T$ and the heat flux q_x, respectively (see Fig. 2.5).

$$\left.\begin{aligned} \mathrm{A} &= \partial(\rho c_p T \Delta x \Delta y \Delta z)/\partial t \\ \mathrm{B} &= (v_x \rho c_p T + q_x)(\Delta y \Delta z) \\ \mathrm{C} &= \{(v_x + \Delta v_x)\rho c_p (T + \Delta T) + (q_x + \Delta q_x)\}(\Delta y \Delta z) \\ \mathrm{D} &= Q\, \Delta x \Delta y \Delta z \end{aligned}\right\}, \tag{2.15}$$

where Q in the D term is the heat generation per unit volume and has the dimension of (J/m^3). The D term is evaluated by taking account of the heat generation in the

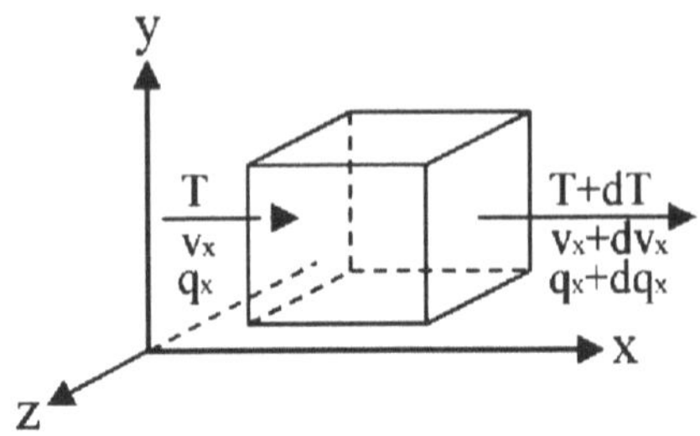

Fig. 2.5 Heat balance in one dimension

small volume element, which is induced by reaction or induction heating, etc.. And Δq_x can be expressed as follows:

$$\Delta q_x = \{\partial(-\lambda\, \partial T/\partial x)/\partial x\}\Delta x = -\lambda(\partial^2 T/\partial x^2)\Delta x$$

This term is substituted into Eq. 2.4 to get Eq. 2.16.

$$\underbrace{c_p\rho(\partial T/\partial t)\Delta x\Delta y\Delta z}_{A} = \underbrace{(v_x\rho c_p T + q_x)(\Delta y\Delta z)}_{B} - \underbrace{[\{v_x + (\partial v_x/\partial x)\Delta x\}\rho c_p\{T + (\partial T/\partial x)\Delta x\} + \{q_x - \lambda(\partial^2 T/\partial x^2)\Delta x\}](\Delta y\Delta z)}_{C} + \underbrace{Q\Delta x\Delta y\Delta z}_{D} \tag{2.16}$$

Dividing both sides of Eq. 2.16 by $\Delta x\Delta y\Delta z$ and rearranging by use of Eq. 2.8 yields Eq. 2.17 for the heat balance in the one-dimensional form.

$$c_p\rho\{\partial T/\partial t + v_x(\partial T/\partial x)\} = \lambda(\partial^2 T/\partial x^2) + Q \tag{2.17}$$

By taking the same procedure as the derivation of Eqs. 2.14 and 2.18 as the equation of heat balance in a three-dimensional formula can be derived.

$$\begin{aligned} &c_p\rho\{\partial T/\partial t + v_x(\partial T/\partial x) + v_y(\partial T/\partial y) + v_z(\partial T/\partial z)\} \\ &\quad = \lambda(\partial^2 T/\partial x^2 + \partial^2 T/\partial y^2 + \partial^2 T/\partial z^2) + Q \\ &or \quad DT/Dt \equiv \partial T/\partial t + (\boldsymbol{v}\cdot\nabla)T = \kappa\nabla^2 T + Q/c_p\rho, \end{aligned} \tag{2.18}$$

where $\kappa \equiv (\lambda/c_p\rho)$ is the thermal diffusivity, which has the same dimension (m^2/s) as the diffusion coefficient of a species D.

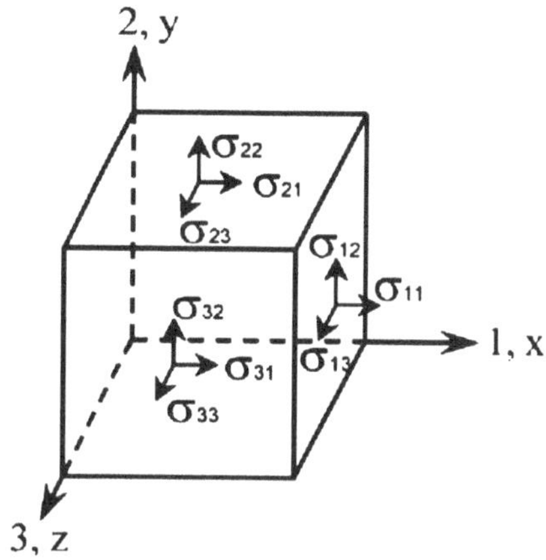

Fig. 2.6 The plane and direction in which stress tensor σ_{ij} acts

2.3 The Equation of Motion

Concentration and temperature are scalar quantities and velocity is a vector. The stress (the force acting on a unit area) acting on a plane is divided into three components which act on the three different planes perpendicular to x, y, and z axes in a small element $\Delta x \Delta y \Delta z$ as shown in Fig. 2.6, in which the suffixes 1, 2 and 3 denote the x, y and z axes, respectively. Thus, there exist nine different stress components which are called (second-order) tensors and then may be written as:

$$\sigma_{i,j} = \begin{pmatrix} \sigma_{11} & \sigma_{12} & \sigma_{13} \\ \sigma_{21} & \sigma_{22} & \sigma_{23} \\ \sigma_{31} & \sigma_{32} & \sigma_{33} \end{pmatrix} \tag{2.19}$$

The lower subscript of σ, i and j express the plane and the direction on which the stress σ acts, respectively.

Following the derivation of the three balance equations mentioned before, let us take the momentum balance in the x-direction. At first, the A term is the time variation of the momentum in a small volume element $\Delta x \Delta y \Delta z$ in the x-direction. It is given in Eq. 2.20.

$$\mathrm{A} = \partial(\rho \Delta x \Delta y \Delta z v_x)/\partial t \tag{2.20}$$

Momentum flows into and out of the volume element by two mechanisms: by convection (that is, by *virtu*e of the bulk flow) and by molecular transfer (that is, by *virtu*e of velocity gradients). The B term is the sum of both the molecular transfer and convection in the x-direction flowing into the element, and can be evaluated by referring Fig. 2.7 as shown in Eq. 2.21.

$$\begin{aligned} \mathrm{B} = {} & \{\sigma_{11}|_{x+\Delta x} + \rho v_x^2|_x\}(\Delta y \Delta z) + \{\sigma_{21}|_{y+\Delta y} + \rho v_x v_y|_y\}(\Delta z \Delta x) \\ & + \{\sigma_{31}|_{z+\Delta z} + \rho v_x v_z|_z\}(\Delta x \Delta y) \end{aligned} \tag{2.21}$$

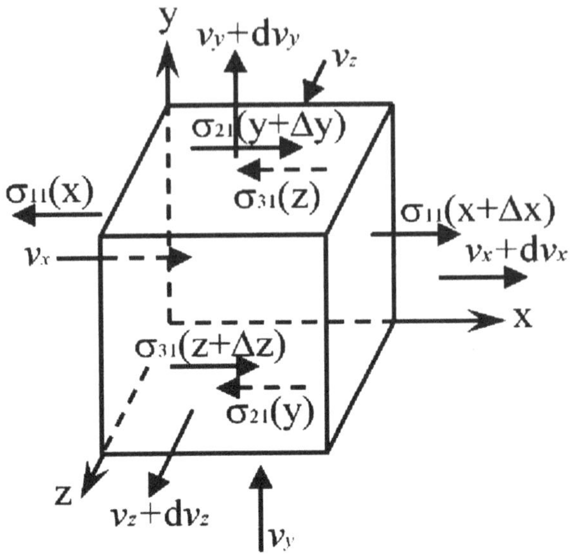

Fig. 2.7 Momentum balance in one dimension

The C term is the sum of both the molecular transfer and convection in the x-direction flowing out from the element. It is expressed as Eq. 2.22.

$$\begin{aligned} \mathrm{C} = &\{\sigma_{11}|_x + \rho v_x^2|_{x+\Delta x}\}(\Delta y\Delta z) + \{\sigma_{21}|_y + \rho v_x v_y|_{y+\Delta y}\}(\Delta z\Delta x) \\ &+ \{\sigma_{31}|_z + \rho v_x v_z|_{z+\Delta z}\}(\Delta x\Delta y) \end{aligned} \tag{2.22}$$

The D term is a body force acting on the element such as gravity force, electromagnetic force, centrifugal force, etc., which are proportional to the volume of the element. The surface tension force is not the body force so that it is not included in D.

$$\mathrm{D} = f_x(\Delta x\Delta y\Delta z), \tag{2.23}$$

where f_x expresses the x-direction body force.

Substituting A ~ D into Eq. 2.4 and rearranging the resulting equation yield Eq. 2.24.

$$\begin{aligned} \{\partial(\rho v_x)/\partial t\}(\Delta x\Delta y\Delta z) = &\{\sigma_{11}|_{x+\Delta x} - \sigma_{11}|_x\}(\Delta y\Delta z) \\ &+ \{\sigma_{21}|_{y+\Delta y} - \sigma_{21}|_y\}(\Delta z\Delta x) \\ &+ \{\sigma_{31}|_{z+\Delta z} - \sigma_{31}|_z\}(\Delta x\Delta y) \\ &+ \{\rho v_x^2|_x - \rho v_x^2|_{x+\Delta x}\}(\Delta y\Delta z) \\ &+ \{\rho v_x v_y|_y - \rho v_x v_y|_{y+\Delta y}\}(\Delta z\Delta x) \\ &+ \{\rho v_x v_z|_z - \rho v_x v_z|_{z+\Delta z}\}(\Delta x\Delta y) + f_x(\Delta x\Delta y\Delta z) \end{aligned} \tag{2.24}$$

By dividing the above equation by $\Delta x \Delta y \Delta z$ and taking the limit as $\Delta x, \Delta y$ and Δz approach zero, we obtain the following equation by using the continuity equation of Eq. 2.10.

$$\rho(\partial v_x/\partial t + v_x\,\partial v_x/\partial x + v_y\,\partial v_x/\partial y + v_z \partial v_x/\partial z) \\ = \partial\sigma_{11}/\partial x + \partial\sigma_{21}/\partial y + \partial\sigma_{31}/\partial z + f_x, \tag{2.25}$$

where the stress σ_{ij} can be expressed for incompressible and Newtonian fluid as:

$$\sigma_{ij} = -p\delta_{ij} + \mu(\partial v_i/\partial n_j + \partial v_j/\partial n_i), \tag{2.26}$$

where δ_{ij} is the Kronecker delta, which takes the value of $\delta_{ij} = 1$ at $i = j$ and $\delta_{ij} = 0$ at $i \neq j$. And n_i $(i = 1,2,3)$ express x, y and z and v_i $(i = 1,2,3)$ do v_x, v_y and v_z, respectively. Thus, σ_{11}, σ_{21} and σ_{31} are written as follows:

$$\left.\begin{aligned} \sigma_{11} &= -\,p + \mu(\partial v_1/\partial n_1 + \partial v_1/\partial n_1) = -p + 2\mu\,\partial v_x/\partial x \\ \sigma_{21} &= \mu(\partial v_2/\partial n_1 + \partial v_1/\partial n_2) = \mu(\partial v_y/\partial x + \partial v_x/\partial y) \\ \sigma_{31} &= \mu(\partial v_3/\partial n_1 + \partial v_1/\partial n_3) = \mu(\partial v_z/\partial x + \partial v_x/\partial z) \end{aligned}\right\} \tag{2.27}$$

Then, Eq. 2.28 is obtained.

$$\begin{aligned} \partial\sigma_{11}/\partial x + \partial\sigma_{21}/\partial y + \partial\sigma_{31}/\partial z &= -\partial p/\partial x + \mu(\partial^2 v_x/\partial x^2 + \partial^2 v_x/\partial y^2 + \partial^2 v_x/\partial z^2) \\ &\quad + \mu(\partial^2 v_x/\partial x^2 + \partial^2 v_y/\partial x\partial y + \partial^2 v_z/\partial x\partial z) \\ &= -\partial p/\partial x + \mu(\partial^2 v_x/\partial x^2 + \partial^2 v_x/\partial y^2 + \partial^2 v_x/\partial z^2) \\ &\quad + \mu\,\partial(\partial v_x/\partial x + \partial v_y/\partial y + \partial v_z/\partial z)/\partial x \end{aligned} \tag{2.28}$$

Since from the equation of continuity, the last term in Eq. 2.28 is zero, we obtain Eq. 2.29 as the momentum equation for incompressible fluid in the x-direction.

$$\rho(\partial v_x/\partial t + v_x \partial v_x/\partial x + v_y \partial v_x/\partial y + v_z \partial v_x/\partial z) \\ = -\partial p/\partial x + \mu(\partial^2 v_x/\partial x^2 + \partial^2 v_x/\partial y^2 + \partial^2 v_x/\partial z^2) + f_x \tag{2.29}$$

By taking the same procedure, the momentum equations for the y- and z-directions are derived as Eqs. 2.30 and 2.31, respectively.

$$\rho(\partial v_y/\partial t + v_x \partial v_y/\partial x + v_y \partial v_y/\partial y + v_z\,\partial v_y/\partial z) \\ = -\partial p/\partial y + \mu(\partial^2 v_y/\partial x^2 + \partial^2 v_y/\partial y^2 + \partial^2 v_y/\partial z^2) + f_y \tag{2.30}$$

$$\rho(\partial v_z/\partial t + v_x \partial v_z/\partial x + v_y\, \partial v_z/\partial y + v_z\, \partial v_z/\partial z)$$
$$= -\partial p/\partial z + \mu(\partial^2 v_z/\partial x^2 + \partial^2 v_z/\partial y^2 + \partial^2 v_z/\partial z^2) + f_z \tag{2.31}$$

By using vector-tensor notation, Eqs. 2.29, 2.30, and 2.31 can be written as follows:

$$\rho\{\partial \boldsymbol{v}/\partial t \quad + \quad (\boldsymbol{v}\cdot\nabla)\boldsymbol{v}\} \quad = \quad -\nabla_p \quad + \quad \mu\nabla^2\boldsymbol{v} \quad + \quad f,$$

Unsteady state term — Convection term — Pressure term — Viscosity term — Force term

(Unsteady state term + Convection term) → Inertial term $\rho(D\boldsymbol{v}/Dt)$

(2.32)

where $\boldsymbol{f}$ is a vector expression given by $\boldsymbol{f} \equiv (f_x, f_y, f_z)$.

This is the well-known *Navier-Stokes equation*, first developed by *Navier, a scientist, and Stokes, a civil engineer*, and it only holds under *constant* ρ *and* μ as you have understood from its derivation.

The more general formula of the equation of motion in which the restriction of constant ρ and μ is released can be easily estimated from Eq. 2.32 as:

$$\rho(D\boldsymbol{v}/Dt) = -\nabla p - [\nabla\cdot\boldsymbol{\sigma}] + \boldsymbol{f}, \tag{2.32$'$}$$

where $D\boldsymbol{v}/Dt \equiv \partial\boldsymbol{v}/\partial t + (\boldsymbol{v}\cdot\nabla)\boldsymbol{v}$ is.

The formulations of Eq. 2.32′ corresponding to rectangular (Cartesian), cylindrical and spherical coordinates are given in reference [3].

The variables appearing in Eqs. 2.29, 2.30, and 2.31 are v_x,v_y,v_z and p, the number of which are four, and the number of the governing equations are three. Thus, we need one more equation to determine all of the variables. By adding the equation of continuity of Eq. 2.10, the number of variables becomes equal to that of the equations. That is, we can determine all of the variables in the velocity field. From the comparison of Eq. 2.32 with the equation of mass balance Eq. 2.14 and the equation of heat balance Eq. 2.18, we find the difference between them in the convection term. That is, on the one hand the convection term of $\rho(\boldsymbol{v}\cdot\nabla)\boldsymbol{v}$ in Eq. 2.32 is in nonlinear form, on the other hand the term of $(\boldsymbol{v}\cdot\nabla)c$ in Eq. 2.14 and the term of$(\boldsymbol{v}\cdot\nabla)T$ in Eq. 2.18 are in linear form. Thus, Eqs. 2.14 and 2.18 are linear equations and Eq. 2.32 is a nonlinear one, so that solving the latter equation is much more difficult in comparison with the former ones.

Equation 2.32 can be applied to describe a laminar flow, but not a turbulent one, where momentum transfer can no longer be described in terms of the velocity gradients and the Newtonian definition of viscosity but rather much more complex and yet less satisfactory relationships have to be invoked for relating the momentum flux to the local velocity gradient. It follows that turbulent flow phenomena are much more complex than laminar flow behavior; moreover, turbulence phenomena are much less well understood. Nonetheless, turbulent flow is encountered in the majority of materials processing operations. The discussion of theoretical descriptions of the turbulence phenomenon and its computational method are beyond the area covered by this book so that readers are recommended to consult other references for turbulent flow [4], turbulent phenomena in MHD flow [5, 6] and mathematical modeling of MHD flow [7].

The governing equations have been derived in rectangular (Cartesian) coordinates, hitherto, and it is possible to derive the same governing equations in the other coordinate systems though its procedure could be more complicated. The governing equations, that is, the equation of continuity: Eq. 2.10, the equation of mass balance: Eq. 2.14, the equation of heat balance: Eq. 2.18 and the equation of momentum balance: Eq. 2.32 are formulated in the selected rectangular (Cartesian), circler cylindrical and spherical coordinates in Appendixes 1–4, respectively, which are shown at the end of this book.

Problem II-2. Derive Eq. 2.30 by following the deriving procedure of Eq. 2.29.

3 Dimensionless Numbers Appearing in Transport Phenomena

In the governing equations of Eqs. 2.14, 2.18 and 2.32, which were derived in the previous section, each term in an equation should have the same dimension so that dimensionless numbers are obtained by taking the ratio of each term. In order to systematically derive the dimensionless numbers appearing in *Transport Phenomena*, the governing equations and their corresponding boundary conditions are enumerated in the following.

Governing Equation **Boundary Condition**

The equation of mass balance

$\partial c/\partial t$	+	$(\boldsymbol{v}\cdot\nabla)c$	=	$D\nabla^2 c$	+	R	$J = k(c_s - c_b)$
unsteady state term		convection term		diffusion term		reaction term	,
(A − I)		(A − II)		(A − III)		(A − IV)	(A − V)

where c_s and c_b are the concentrations on a surface and in bulk, respectively.

The equation of heat balance

$c_p\rho\{\partial T/\partial t$	+	$(\boldsymbol{v}\cdot\nabla)T\}$	=	$\lambda\nabla^2 T$	+	Q	$q = h(T_s - T_b)$
unsteady state term		convection term		diffusion term		heat generation term	.
(B − I)		(B − II)		(B − III)		(B − IV)	(B − V)

where T_s and T_b are the temperatures on a surface and in bulk, respectively, and h is the heat transfer coefficient

The equation of momentum balance

$\rho\{\partial\boldsymbol{v}/\partial t$	+	$(\boldsymbol{v}\cdot\nabla)\boldsymbol{v}\}$	=	$\mu\nabla^2\boldsymbol{v}$	+	f	+	$\nabla p,$	τ_0=const., $\tau_0=\sigma_f/L$
unsteady state term		convection term		viscosity term		force term		pressure term	$\nabla\tau_0$ = const.,
(C − I)		(C − II)		(C − III)		(C − IV)		(C − V)	(C − VI)

where σ_f is the surface tension, and the equation $\tau_0 = \sigma_f/L$ is a boundary condition for the surface of a fluid having curvature, and the equations of τ_0 = const. and $\nabla\tau_0$ = const. express the boundary conditions for the fluid in contact with a solid and another fluid, respectively.

By taking the ratio of each term, we can obtain the following dimensionless numbers.

Dimension		Dimensionless number
(A-II)/(A-III) $= (cV/L)/(Dc/L^2)$	$= LV/D$	*Pe': mass Peclet number*
(A-V)/(A-III) $= (kc/L)/(Dc/L^2)$	$= kL/D$	*Sh: Sherwood number*
(A-IV)/(A-II) $= (R)/(cV/L)$	$= RL/cV$	D_I*: Damköhler-I number*
(A-IV)/(A-III) $= (R)/(Dc/L^2)$	$= RL^2/Dc$	D_{II}*: Damköhler-II number*
(B-II)/(B-III) $= (c_p\rho TV/L)/(\lambda T/L^2)$	$= Vc_p\rho L/\lambda$	*Pe: heat Peclet number*
(B-IV)/(B-II) $= (Q)/(c_p\rho VT/L$	$= QL/c_p\rho VT$	D_{III}*: Damköhler-III number*
(B-III)/(B-I) $= (\lambda T/L^2)/(c_p\rho T/t)$	$= \lambda t/c_p\rho L^2$	*Fo: Fourier number*
(B-VI)/(B-III) $= (hT/L)/(\lambda T/L^2)$	$=hL/\lambda$	*Nu: Nusselt number*
(C-II)/(C-III) $= (\rho V^2/L)/(\mu V/L^2)$	$= LV\rho/\mu$	*Re: Reynolds number*
(C-II)/(C-IV) $= (\rho V^2/L)/(\rho g)$	$= V^2/gL$	*Fr: Froude number*
(C-IV)/(C-II) $= (\sigma B^2 V)/(\rho V^2/L)$	$=\sigma B^2 L/\rho V$	*N:Stuart number*
(C-V)/(C-II) $= (\Delta P/L)/(\rho V^2/L)$	$= \Delta P/\rho V^2$	*f: Friction Factor*
(C-VI)/(C-III) $= (\tau_0/L)/(\mu V/L^2)$	$=\tau_0 L/\mu V$	*Bi: Bingham number*
(C-II)/(C-VI) $= (\rho V^2/L)/(\sigma_f/L^2)$	$= \rho V^2 L/\sigma_f$	*We: Weber number*
(C-IV)/(C-III) $= (\sigma B^2 V)/(\mu V/L^2)$	$= B^2 L^2\sigma/\mu$	*Ha: Hartmann number*

Furthermore, the following dimensionless numbers are obtained by overlapping of equations and the dimensionless numbers.

$$\textit{Grashof number}: \; Gr = (\beta g\rho^2 L^3\Delta T/\mu^2) = (L^2V^2\rho^2/\mu^2)\cdot(gL/V^2)\cdot(\beta\Delta T)$$
$$= Re^2\cdot(1/Fr)\cdot(\Delta\rho/\rho),$$

where β is the thermal coefficient of volumetric expansion ($1/K$).

Yakob number: $Y = c_p\Delta T/\Delta H$, where ΔH is the enthalpy change (*J/kg*).

Prandtl number $Pr = (\mu/\rho)/(\lambda/c_p\rho) = \nu/\kappa$,

where ν is the *kinematic viscosity* defined by $\nu \equiv \mu/\rho$ and κ is the *thermal diffusivity* defined as $\kappa \equiv \lambda/c_p\,\rho$.

Rayleigh number: $Ra = Gr\cdot Pr$

Schmidt number: $Sc = Pe'/Re = \nu/D$

Lewis number: $Le = Sc/Pr = \kappa/D$

Biot number: $Bo = hL/\lambda$,

where the *Biot number Bo* is the same formula with the *Nusselt number*, but the difference is that λ in the *Biot number* indicates the *thermal conductivity* $\{J/(m\,s\,K)\}$ in a solid and the *Nusselt number* does so in a fluid.

4 The Procedure for Solving Problems by Use of Governing Equations

Here, let us show the procedure of solving problems by use of the governing equations given in the previous section. You will find how to solve them in examples. The procedure is composed of the following three steps:

(I) The coordinate suitable for given boundary conditions is selected.
(II) The unnecessary terms in the governing equations are eliminated.
(III) The governing equation is solved under the given boundary and initial conditions.

4.1 *The Mass Diffusion Model in the Infinite One-Dimensional Field*

Let us image an infinite one-dimensional field (which is composed of a media combining gas, liquid and solid, and let us pick up the liquid media, here) and assume that the liquid phase comes into contact with the gas phase at the interface of $z=0$ and the concentration of the liquid at $z=0$ is kept at the equilibrium concentration c_s. In the case where a molten metal comes into contact with a gas, the reaction rate is fast enough to keep a chemical equilibrium at the interface in comparison with such process variations as interface shape change and rising velocity of bubbles. Thus, this assumption holds in most cases that we face in industry. Let us assume that no fluid motion exists and the diffusion takes place in the z-direction only. The concentration variation in the liquid with time is schematically shown as Fig. 2.8. Let us solve this problem along the procedure of (I)–(III) given above.

(I) **The coordinate suitable for given boundary conditions is selected.**
The rectangular (Cartesian) coordinate is selected as the simplest one.

(II) **The unnecessary terms in the governing equations are eliminated.**
The equation to be solved is Eq. A.4 given in Appendix 2. Since it is assumed that the velocity of a liquid is zero, namely $\boldsymbol{v} = (v_x, v_y, v_z) = 0$, and diffusion

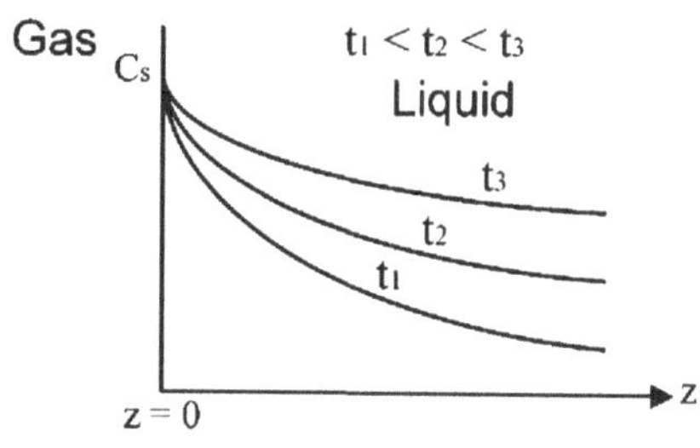

Fig. 2.8 Schematic view showing the change in concentration profile with time

does not take place in the x- and y-directions, the governing equation is simplified as Eq. 2.33.

$$\partial c/\partial t = D\,\partial^2 c/\partial z^2 \tag{2.33}$$

(III) **The governing equation is solved under the given boundary and initial conditions.**

The boundary conditions:

$$c = c_s \text{ at } z = 0,\ \ t \geq 0 \ \ \text{(the gas – liquid equilibrium at theinterface)} \tag{2.34}$$

$$c = c_b \text{ at } z = \infty,\ t \geq 0, \tag{2.35}$$

where Eq. 2.35 implies that the concentration in a bulk liquid located far from the interface is kept at c_b . As will be shown, the diffusion length of species l is $l = \sqrt{Dt}$. Putting a value of the diffusion coefficient of a liquid at about 5×10^{-9} m^2/s gives $l = 0.22$ mm at $t = 10$ s and $l = 0.7$ mm at $t = 100$ s. If we would consider a several times longer distance than these values, the obtained concentration profile could be almost the same as that calculated under the condition of $c = c_b$ at $z = \infty$.

The initial condition:

$$c = c_b \text{ at } t = 0,\ \ z > 0, \tag{2.36}$$

where Eq. 2.36 means that the concentration in the entire area is kept at $c = c_b$ at first.

Let us solve Eq. 2.33 under the boundary and initial conditions of Eqs. 2.34 and 2.35 and Eq. 2.36, respectively. Let us put $c' \equiv c - c_b$ so as to simplify the boundary and initial conditions.

The governing equation:

$$\partial c'/\partial t = D\partial^2 c'/\partial z^2 \tag{2.37}$$

The boundary conditions:

$$c' = c_s - c_b \text{ at } \ z = 0,\ \ t \geq 0 \tag{2.38}$$

$$c' = 0 \quad \text{at } z = \infty,\ t \geq 0 \tag{2.39}$$

The initial condition:

$$c' = 0 \quad \text{at } t = 0,\ \ z > 0 \tag{2.40}$$

The solution is given in Eq. 2.41 (The precise derivation is given in Appendix A)

$$c' = c - c_0 = (c_s - c_b)\{1 - erf(z/2\sqrt{Dt})\} \tag{2.41}$$

What we want to get is not the concentration profile, but how much gas is absorbed across the interface. Thus, the mass flux defined in Eq. 2.1 is obtained as follows:

$$J_z = -D(\partial c/\partial z) = \sqrt{D/\pi t}(c_s - c_b)\ exp\{-(z/2\sqrt{Dt})^2\} \tag{2.42}$$

The mass flux across the interface is given by putting z = 0 in Eq. 2.42.

$$J_z = \sqrt{D/\pi t}(c_s - c_b) \tag{2.43}$$

Equation 2.43 indicates the instantaneous rate of the mass flux. Then, the instantaneous (at any moment) mass transfer coefficient k' is defined as Eq. 2.44.

$$k' = \sqrt{D/\pi t} \tag{2.44}$$

The averaged absorption rate over $0 \sim t_e$ is obtained as Eq. 2.45.

$$\bar{J_z} = (1/t_e)\int_0^{t_e} \sqrt{D/\pi t}(c_s - c_b)dt = 2\sqrt{D/\pi t_e}(c_s - c_b) \tag{2.45}$$

Then, the averaged mass transfer coefficient *k*, which is simply called the mass transfer coefficient in general, is defined as Eq. 2.46.

$$k = 2\sqrt{D/\pi t_e}, \tag{2.46}$$

where t_e is the contacting time of gas and liquid.

Let us now explain the meaning of introducing the mass transfer coefficient *k*. When we design a process as large as *0.01–10 m*, we cannot take account of the behavior of each atom and molecule having nanometer size of $10^{-9}m$ since a big size difference, as much as $10^{9\text{–}10}$, exists between the process and the atoms. This scale difference can be imagined as follows: If the size of the atoms was 1 mm, the size of the process could have been *1,000–10,000 km*. When we have to consider matters with such a big scale difference, the information in the under level (the smaller scale) has to be concentrated in a lumped value which is taken into the upper level (the larger scale). The concept of the mass transfer coefficient is a good example which is introduced so as to lump the phenomena occurring in the atomic size. This concept of the procedure is called "*layering*", which is an important concept in the field of *Chemical Engineering*.

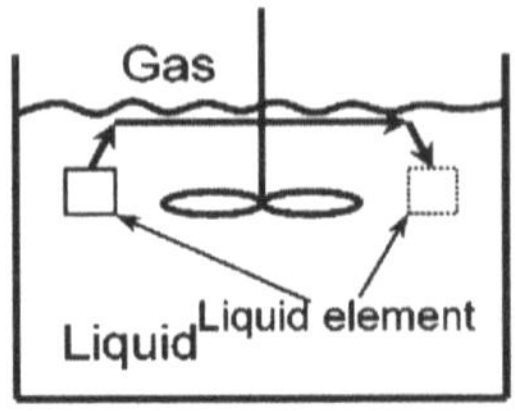

Fig. 2.9 Conceptual figure of gas absorption in a liquid under complete mixing

Now, let us discuss how to use the instantaneous coefficient k' and the averaged mass transfer coefficient k. Let us take the carburizing or nitrifying process as examples, where carbon or nitrogen atoms are diffused into a solid piece submerged in the corresponding atomic gas or liquid phase. The diffusion rate of atom into the solid piece can be obtained by use of the instantaneous mass transfer coefficient k' as Eq. 2.47.

$$w = k'(c_s - c_b) \times A, \tag{2.47}$$

where A is the surface area of the solid and w the mass transfer rates.

On the other hand, in the case where gas absorption takes place from the surface of a liquid stirred in a vessel, as shown in Fig. 2.9, we can imagine that a small liquid element starts to absorb gas at the moment when the liquid element appears on the surface, and stops it when the liquid element is transported from the surface to a bulk. The time interval t_e from appearance of the liquid element on the surface to disappearance is called an averaged contacting time, which is tightly connected with the intensity of stirring in the vessel. In the case where the intensity is strong enough to keep the concentration of the bulk liquid in the vessel homogeneous, the concentration is expressed by use of the averaged mass transfer coefficient k as Eq. 2.48.

$$Vdc_b/dt = Ak(c_s - c_b), \tag{2.48}$$

where V is the volume of the liquid in a vessel.

As the initial condition is given as follows:

$$\mathrm{c} = c_0 \quad \text{at } t = 0 \tag{2.49}$$

the solution is obtained as Eq. 2.50.

$$c_b = c_s - (c_s - c_0) \cdot exp(-Akt/V) \tag{2.50}$$

The Eq. 2.50 expresses the change in concentration of the liquid in the vessel with time.

Problem II-3. Derive Eq. 2.42.

Problem II-4. Derive Eq. 2.50.

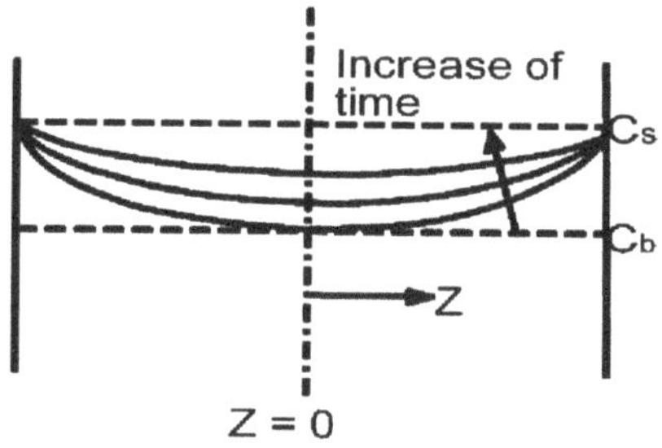

Fig. 2.10 Concentration profile with time in a flat plate

Problem II-5. Describe the difference between c_b in Eq. 2.35 and c_0 in Eq. 2.49.

4.2 *The Diffusion Model in the Finite One-Dimensional Field*

4.2.1 The Case of a Flat Plate

Let us take a flat plate having a thickness of $2l$ in one direction and a large length in the other two directions as a diffusion model, as shown in Fig. 2.10. The diffusion only takes place in the z-direction, namely in the thickness direction. The governing equation, the boundary and initial conditions for this problem are given as follows:
The Governing Equation:

$$\partial c/\partial t = D\partial^2 c/\partial z^2 \tag{2.33}$$

The Boundary Conditions:

$$\partial c/\partial z = 0 \text{ at } z = 0, \quad t > 0 \tag{2.51}$$

$$c = c_s \quad \text{at } z = \pm\, l, \quad t > 0 \tag{2.52}$$

The Initial Condition:

$$c = c_b \qquad \text{at } t = 0, \quad |z| < l \tag{2.53}$$

To get the solution, let us take the two methods of Fourier transform and Laplace transform.

(1) The solution obtained by the Fourier transform (the derivation of the solution is given in Appendix B)

$$c = c_s + 4\{(c_b - c_s)/\pi\} \sum_{m=0}^{\infty} \{1/(2m+1)\}(-1)^m \cdot$$
$$exp\{-D(2m+1)^2\pi^2 t/4l^2\} \times \cos\{(2m+1)\pi z/2l\} \tag{2.54}$$

The instantaneous and averaged mass transfer coefficients k' and k are obtained as follows:

$$k' = (2D/l) \sum_{m=0}^{\infty} exp\ \{-D\,(2m+1)^2\pi^2 t/4l^2\} \tag{2.55}$$

$$k = (8l/t_e\pi^2) \sum_{m=0}^{\infty} [1 - exp\{-D(2m+1)^2\pi^2 t/4l^2\}]/(2m+1)^2 \tag{2.56}$$

(2) The solution obtained by the Laplace transform (the derivation of the solution is given in Appendix C)

$$c = c_b + (c_s - c_b)\left[\sum_{n=0}^{\infty}(-1)^n erfc\left\{\frac{(2n+1)l + z}{2\sqrt{Dt}}\right\} + \sum_{n=0}^{\infty}(-1)^n erfc\left\{\frac{(2n+1)l - z}{2\sqrt{Dt}}\right\}\right] \tag{2.57}$$

In this case, the instantaneous mass transfer coefficient k' is obtained as shown in Eq. 2.58, but the averaged mass transfer coefficient cannot be expressed in the elementary functions so that it is not given here.

$$k' = \sqrt{D/\pi t}\left\{1 + 2\sum_{n=1}^{\infty}(-1)^n exp(-n^2 l^2/Dt)\right\} \tag{2.58}$$

Now, let us distinguish each use of the solutions of the Fourier transform and the Laplace transform. Each concentration variation at $z = 0$, with elapsed time, which is given as Eqs. 2.59 and 2.60 are shown in Fig. 2.11.

$$(c - c_b)/(c_s - c_b) = 1 - \frac{4}{\pi}\sum_{m=0}^{\infty}\frac{(-1)^m}{2m+1} exp\left\{\frac{-D(2m+1)^2\pi^2 t}{4l^2}\right\} \quad \text{(Fourier transform)} \tag{2.59}$$

$$(c - c_b)/(c_s - c_b) = 2\sum_{n=0}^{\infty}(-1)^n\left[1 - erf\left\{\frac{(2n+1)l}{2\sqrt{Dt}}\right\}\right] \quad \text{(Laplace transform)} \tag{2.60}$$

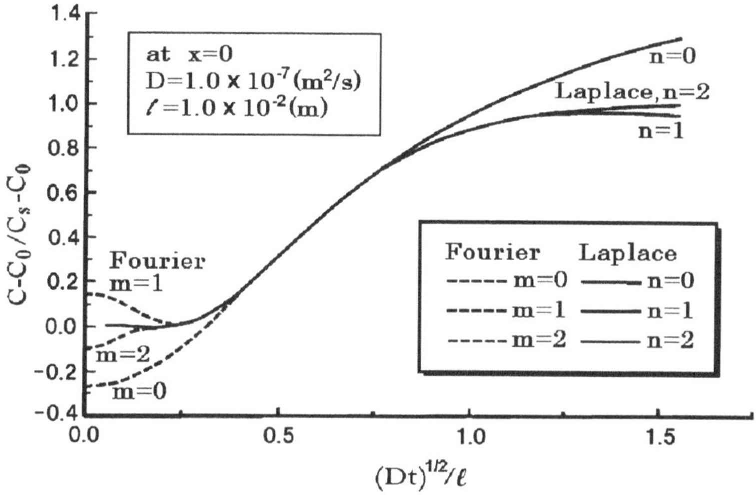

Fig. 2.11 Convergence in the Laplace and Fourier solutions

The solutions approximated by only the first term in the series solution are shown by $m = 0$ for Fourier and $n = 0$ for Laplace, respectively, and the solutions taking account till the second term are $m = 1$ and $n = 1$, and the ones till the third term $m = 2$ and $n = 2$. From this figure, we find the following;

1. For $\sqrt{Dt}/l < 0.6$, the convergence is better in the Laplace solution than in the Fourier one. The accuracy of the solution is fully confirmed by the Laplace solution for $n = 0$.
2. *For* $\sqrt{Dt}/l > 0.6$, the solution by the Fourier transformation is better and its accuracy is enough to take the first term ($m = 0$).

That is, we can recognize that the Laplace solution of Eq. 2.60 is better for the short time range and the Fourier one of Eq. 2.59 for the long time range.

4.2.2 The Case of a Cylindrical Bar

Let us take the case where a long and cylindrical steel bar is set in a gas atmosphere. It is assumed that the gas diffusion into the bar takes place only in the radial direction of the bar by neglecting fringing effect at the edge of the bar. On the basis of the coordinates shown in Fig. 2.12, the governing equation, and the boundary and initial conditions are written as follows:

The governing equation:

$$\partial c/\partial t = D\ \{\partial^2 c/\partial r^2 + (1/r)\ \partial c/\partial r\} \tag{2.61}$$

The boundary conditions:

$$\partial c/\partial r = 0 \quad \text{at} \quad r = 0, \quad t \geq 0 \quad \text{(symmetrical concentration distribution)} \tag{2.62}$$

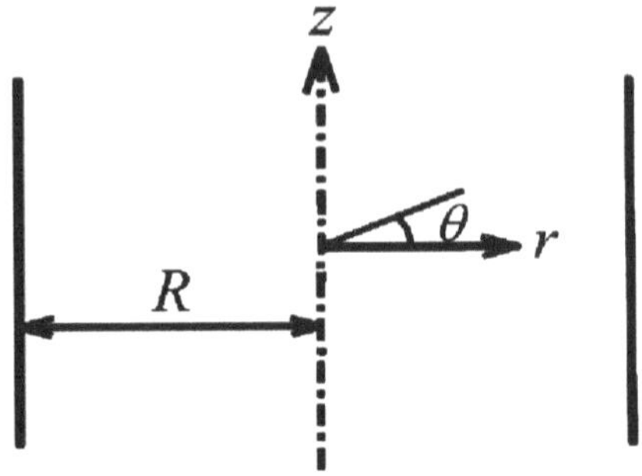

Fig. 2.12 Cylindrical coordinates system

$$c = c_s \qquad \text{at } r = R,\ t \geq 0 \tag{2.63}$$

The initial condition:

$$c = c_b \qquad \text{at } 0 \leq r < R,\ t = 0 \tag{2.64}$$

The concentration profile is obtained by solving this problem (the detail derivation of the solution is given in Appendix D)

$$c = \{2\,(c_b - c_S)/R\} \sum_{n=1}^{\infty} [J_0(\lambda_n r)/\{\lambda_n J_1(\lambda_n R)\}] \cdot exp\left(-D\lambda_n^2 t\right) + c_s, \tag{2.65}$$

where λ_n is the variables which are obtained as the positive root of $J_0(\lambda_n R) = 0$, and J_0 and J_1 are referred to the Bessel function of the first kind of order 0 and 1, respectively. The instantaneous and averaged mass transfer coefficients are obtained from Eq. 2.65, as Eqs. 2.66 and 2.67, respectively.

$$k' = (2D/R) \sum_{n=1}^{\infty} exp\left(-D\lambda_n^2 t\right) \tag{2.66}$$

$$k = (2/t_e R) \sum_{n=1}^{\infty} \left[-\left(1/\lambda_n^2\right)\left\{exp\left(-D\ \lambda_n^2 t_e\right) - 1\right\}\right] \tag{2.67}$$

4.2.3 The Case of a Spherical Ball

Let us imagine that a solid with spherical shape is hung in a chamber filled with gas or liquid. The diffusion of a species in gas or liquid takes place in the solid toward radial direction. On the spherical coordinates system shown in Fig. 2.13, the governing equation and the boundary and initial conditions are shown as follows:

The governing equation:

$$\partial c/\partial t = D\left\{\partial^2 c/\partial r^2 + (2/r)\ (\partial c/\partial r)\right\} \tag{2.68}$$

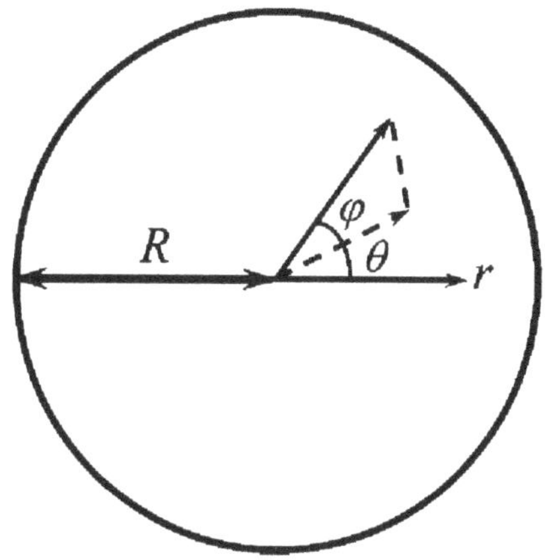

Fig. 2.13 Spherical coordinates system

The boundary conditions:

$$\partial c/\partial r = 0 \text{ at } r = 0, \quad t \geq 0 \text{ (symmetrical concentration distribution)} \tag{2.69}$$

$$c = c_s \qquad \text{at } r = R, \ t \geq 0 \tag{2.70}$$

The initial condition:

$$c = c_b \qquad \text{at} \quad t = 0, \ 0 \leq r < R, \tag{2.71}$$

The solution is derived as Eq. 2.72 (The derivation is given in Appendix E)

$$c = c_s + \frac{2R(c_b - c_s)}{\pi r} \sum_{n=1}^{\infty} \frac{(-1)^{n+1}}{n} \, exp\left(-\frac{n^2\pi^2 D}{R^2} t\right) \cdot sin \frac{n\pi}{R} r \tag{2.72}$$

The instantaneous and averaged mass transfer coefficients are derived as follows:

$$k' = (2D/R) \sum_{n=1}^{\infty} exp \left\{-(n^2\pi^2 D \ t)/R^2\right\} \tag{2.73}$$

$$k = (2R/t_e) \sum_{n=1}^{\infty} (1/n^2\pi^2) \left[1 - exp \left\{-(n^2\pi^2 D \ t_e)/R^2\right\}\right] \tag{2.74}$$

Problem II-6. Derive the time-averaged mass transfer coefficient from Eq. 2.65 in the radial one-dimensional diffusion in a cylindrical coordinates system.

Problem II-7. Derive the time-averaged mass transfer coefficient from Eq. 2.72 in the radial one-dimensional diffusion in a spherical coordinates system.

4.3 The Laminar Flow in a Cylindrical Tube

Let us consider the laminar flow in a cylindrical tube. At first we derive the relations among the velocity distribution, flow rate and tube diameter. The derivation should be followed by the procedure of (I), (II) and (III) proposed in the beginning of this Sect. 4. Of course the cylindrical coordinates system should be chosen in this case. Then, the equation of continuity in the cylindrical coordinates system is copied from Appendix 1 as follows:

$$(1/r)\ \partial\ (rv_r)/\partial r + (1/r)\ \partial v_\theta/\partial\theta + \partial v_z/\partial z = 0 \tag{A.2}$$

Since the flow is laminar with symmetrical pattern about the z-axis, the boundary conditions are given as $v_r = v_\theta = 0$ and $\partial v_z/\partial\theta = 0$. Then, the unnecessary terms are eliminated from (A.2), and finally the following term remains.

$$\partial v_z/\partial z = 0 \tag{2.75}$$

From Appendix 4, the motion equations are given as follows:
The r-component:

$$\rho\left(\frac{\partial v_r}{\partial t} + v_r\frac{\partial v_r}{\partial r} + \frac{v_\theta}{r}\frac{\partial v_r}{\partial\theta} - \frac{v_\theta^2}{r} + v_z\frac{\partial v_r}{\partial z}\right)$$
$$= -\frac{\partial p}{\partial r} + \mu\left[\frac{\partial}{\partial r}\left\{\frac{1}{r}\frac{\partial}{\partial r}(rv_r)\right\} + \frac{1}{r^2}\frac{\partial^2 v_r}{\partial\theta^2} - \frac{2}{r^2}\frac{\partial v_\theta}{\partial\theta} + \frac{\partial^2 v_r}{\partial z^2}\right] + f_r \tag{A.22}$$

Eliminating unnecessary terms from Eq. A.22 yields Eq. 2.76.

$$0 = -\partial p/\partial r \tag{2.76}$$

The θ-component:

$$\rho\left(\frac{\partial v_\theta}{\partial t} + v_r\frac{\partial v_\theta}{\partial r} + \frac{v_\theta}{r}\frac{\partial v_\theta}{\partial\theta} + \frac{v_\theta v_r}{r} + v_z\frac{\partial v_\theta}{\partial z}\right)$$
$$= -\frac{1}{r}\frac{\partial p}{\partial\theta} + \mu\left[\frac{\partial}{\partial r}\left\{\frac{1}{r}\frac{\partial}{\partial r}(rv_\theta)\right\} + \frac{1}{r^2}\frac{\partial^2 v_\theta}{\partial\theta^2} + \frac{2}{r^2}\frac{\partial v_r}{\partial\theta} + \frac{\partial^2 v_\theta}{\partial z^2}\right] + f_\theta \tag{A.23}$$

By eliminating the unnecessary terms, Eq. 2.77 is obtained.

$$0 = -(1/r)(\partial p/\partial\theta) \tag{2.77}$$

The z-component:

$$\rho\left(\frac{\partial v_z}{\partial t}+v_r\frac{\partial v_z}{\partial r}+\frac{v_\theta}{r}\frac{\partial v_z}{\partial \theta}+v_z\frac{\partial v_z}{\partial z}\right)$$
$$=-\frac{\partial p}{\partial z}+\mu\left\{\frac{1}{r}\frac{\partial}{\partial r}\left(r\frac{\partial v_z}{\partial r}\right)+\frac{1}{r^2}\frac{\partial^2 v_z}{\partial \theta^2}+\frac{\partial^2 v_z}{\partial z^2}\right\}+f_z \tag{A.24}$$

Eliminating unnecessary terms yields Eq. 2.78.

$$\underbrace{\rho v_z\frac{\partial v_z}{\partial z}}_{=\,0\text{ from Eq. 2.75}}=-\frac{\partial p}{\partial z}+\mu\left\{\frac{1}{r}\frac{\partial}{\partial r}\left(r\frac{\partial v_z}{\partial r}\right)+\underbrace{\frac{\partial^2 v_z}{\partial z^2}}_{=\,0\text{ from Eq. 2.75}}\right\} \tag{2.78}$$

Then,

$$-\partial p/\partial z+\mu(1/r)\,[\partial\{r\,(\partial v_z/\partial r)\}/\partial r]=0 \tag{2.79}$$

It is recognized that the term of $\partial p/\partial z$ has a possibility to be a function of z, not r and θ as found from Eqs. 2.76 and 2.77, on the other hand the term of $(1/r)[\partial\{r(\partial v_z/\partial r)\}/\partial r]$ has a possibility to be a function of r and θ as understood from Eq. 2.75. Thus, it is found that in Eq. 2.79, the function of z, $\partial p/\partial z$ and the function of r and θ, $(1/r)[\partial\{r(\partial v_z/\partial r)\}/\partial r]$ coexist. This condition is satisfied only when $\partial p/\partial z$ is constant.

$$\partial p/\partial z=\mathit{constant} \tag{2.80}$$

The boundary conditions for p are given as Eqs. 2.81 and 2.82.

$$p=p_0\quad\text{at } z=0 \tag{2.81}$$

$$p=p_L\quad\text{at } z=L \tag{2.82}$$

From Eqs. 2.80, 2.81, and 2.82, p is given as Eq. 2.83.

$$p=p_0+(p_L-p_0)(z/L) \tag{2.83}$$

Thus,

$$\partial p/\partial z=(p_L-p_0)/L \tag{2.84}$$

The boundary conditions for v_z are

$$\partial v_z/\partial r=0\quad\text{at } r=0\quad\text{(symmetrical about z-axis)} \tag{2.85}$$

$$v_z=0\ \text{at } r=R\ \text{(the velocity is zero on the inner wall of the cylindrical tube)} \tag{2.86}$$

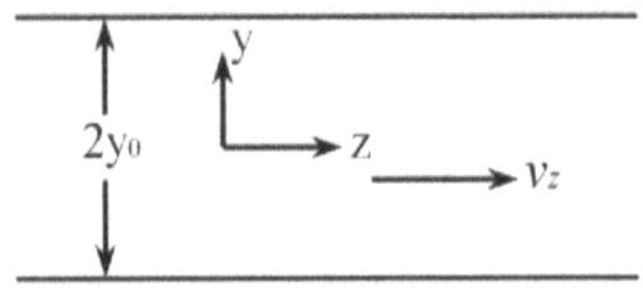

Fig. 2.14 The flow between parallel walls

By solving Eq. 2.79 under the condition of $\partial p/\partial z = \text{constant}$, Eq. 2.87 is obtained.

$$v_z = -\{(p_0 - p_l)/4L\mu\}\ (r^2 - R^2) \tag{2.87}$$

It is found that the velocity profile in the tube is parabolic.

The volumetric flow rate Q is obtained as follows:

$$Q = \int_0^R v_z \cdot 2\pi r dr = \pi(p_L - p_0)R^4/8\mu L \tag{2.88}$$

The above equation giving the relation between the volumetric flow rate Q and the diameter of the tube $2R$ indicates that the volumetric flow rate is in proportion to the fourth power of the diameter, R^4. It suggests that the flow rate is drastically decreased when the diameter becomes narrow due to some reasons. If we would like to keep the flow rate constant under the narrower diameter, the pressure difference between inlet and outlet has to be increased.

4.4 The Laminar Flow Between Parallel Walls

Let us consider a fully developed laminar flow between parallel walls. The distance between the walls is $2\ y_0$ and the coordinates used in this problem are shown in Fig. 2.14. By assuming that the flow does not take place in the x-direction and y-direction, the governing equation and boundary conditions are given as follows:

The governing equation:

$$-\partial p/\partial z + \mu\ (\partial^2 v_z/\partial y^2) = 0 \tag{2.89}$$

The boundary conditions:

$$\left.\begin{array}{ll} p = p_0 & \text{at } z = 0 \\ p = p_L & \text{at } z = L \\ \partial v_z/\partial y = 0 & \text{at } y = 0 \\ v_z = 0 & \text{at } y = y_0 \end{array}\right\} \tag{2.90}$$

The flow pattern is obtained as Eq. 2.91.

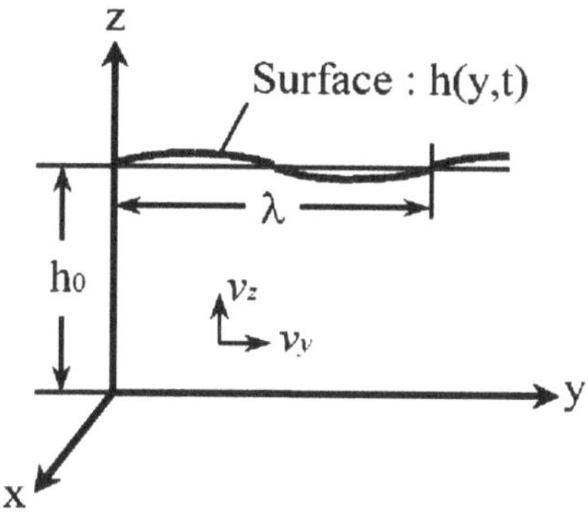

Fig. 2.15 Coordinates and variables in wave motion analysis

$$v_z = \{(p_0 - p_L)/2\mu L\} \cdot y_0^2\{1 - (y/y_0)^2\} \tag{2.91}$$

The volumetric flow rate per unit length in the x-direction is given as the following equation.

$$Q = \{2(p_0 - p_L)/6\mu L\} \cdot y_0^3 \tag{2.92}$$

It is understood that the flow rate Q is in proportion to y_0^3.

Problem II-8. Derive Eqs. 2.91 and 2.92.

4.5 Analysis of Surface Wave Motion

Here, the linear wave theory which treats the waves with small amplitude appearing on the liquid surface is developed. Let us introduce a velocity potential Φ under the assumption of irrotationality, which is defined in Eq. 2.93.

$$\boldsymbol{v} = -\nabla\,\Phi, \quad \boldsymbol{v} = (0,\ v_y,\ v_z) \tag{2.93}$$

Substituting Eq. 2.93 into the equation of continuity Eq. 2.10 yields the following Laplace's equation.

$$\nabla^2\Phi = 0 \tag{2.94}$$

As shown in Fig. 2.15, at the bottom ($z = 0$), the velocity of v_z must be zero so that the boundary condition of Φ is given as follows:

$$\partial\Phi/\partial z = 0 \ \text{at} \ z = 0 \tag{2.95}$$

Now, let us assume the function of Φ as Eq. 2.96 so as to satisfy the Eq. 2.94.

$$\Phi = C\cosh(\xi z) \cdot \cos(\xi y) \cdot exp(-i\omega t) \tag{2.96}$$

By neglecting the viscous and the convection terms on the basis of the linear wave theory and adopting only the gravity force as an external force, the following equation can be obtained from the equation of motion of Eq. 2.32.

$$\rho\, \partial(-\nabla\Phi)/\partial t = -\nabla p + \nabla(-\rho g z), \tag{2.97}$$

where the potential due to the gravity force is expressed as the second term in the right-hand side. Integrating the above equation gives the Bernoulli equation.

$$\rho\, \partial\Phi/\partial t = p + \rho g z \tag{2.98}$$

As $p = 0$ is valid on the free surface, Eq. 2.99 is obtained as the boundary condition on the surface.

$$\rho\, \partial\Phi/\partial t\ |_{z=h} - \rho g z = 0 \tag{2.99}$$

As differentiating Φ in the z-direction at the surface $z = h$ yields $-v_z$, which is the surface velocity with minus sign and corresponds to the change of the free surface position $\partial h/\partial t$, the following equation is obtained.

$$v_z = -\partial\Phi/\partial z\ |_{z=h} = \partial h/\partial t \tag{2.100}$$

Eliminating h by use of Eqs. 2.99 and 2.100, and putting $h \approx h_0$ on the basis of the approximation of a small amplitude wave give Eq. 2.101.

$$\{\partial\Phi/\partial z + (1/g)(\partial^2\Phi/\partial t^2)\}_{z=h_0} = 0 \tag{2.101}$$

Substituting Eq. 2.96 into Eq. 2.101 yields the relation between angular frequency ω and wave number $\xi \equiv 2\pi/\lambda$ as Eq. 2.102, where λ is the wave length.

$$\omega^2 = g\xi \tanh\ (h_0\xi) \tag{2.102}$$

Problem II-9. Derive Eq. 2.97 by neglecting the viscous and convection terms in the *Navier-Stokes* equation of Eq. 2.32 and substituting the gravity force in the force term.

Appendixes A–E

Appendix A: Methods for Solving Eq. 2.37

(1) The method of solving the partial differential equation 2.37 by transforming it to an ordinary differential equation.

Let us assume c' is a function of only the non-dimensional variable $\eta \equiv z/2\sqrt{Dt}$

$$c' = f(\eta) \tag{2.103}$$

Then,

$$\partial\eta/\partial t = -\left(z/4\sqrt{Dt}\right)\ t^{-1} = -(\eta/2t) \tag{2.104}$$

Thus,

$$\partial c'/\partial t = (df/d\eta)(d\eta/dt) = -(\eta/2t)(df/d\eta) \tag{2.105}$$

On the other hand,

$$\partial\eta/\partial t = 1/2\sqrt{Dt} \tag{2.106}$$

is valid, so that

$$\partial^2 c'/\partial z^2 = \frac{\partial}{\partial z}(\partial f/\partial z) = (\partial\eta/\partial z)\frac{d}{d\eta}\left(\frac{\partial\eta}{\partial z}\frac{df}{d\eta}\right) = \frac{1}{4Dt}\frac{d^2 f}{d\eta^2} \tag{2.107}$$

By substituting Eqs. 2.105 and 2.107 into 2.37 and rearranging, the following equation is obtained.

$$d^2f/d\eta^2 + 2\eta\ (df/d\eta) = 0 \tag{2.108}$$

The boundary conditions of Eqs. 2.38 and 2.39 and the initial condition of Eq. 2.40 are written as follows:

The boundary conditions are

$$f = c_s - c_0 \qquad \text{at} \quad \eta = 0 \tag{2.109}$$

$$f = 0 \qquad \text{at} \quad \eta = \infty \tag{2.110}$$

The initial condition is

$$f = 0 \qquad \text{at} \quad \eta = \infty \tag{2.111}$$

Equation 2.108 is an ordinary differential equation so that it is easily integrated twice to get a solution as Eq. 2.112.

$$f = c_1 \int_0^{\eta} exp\left(-\eta^2\right)\ d\eta + c_2 \tag{2.112}$$

The integral constants of c_1 and c_2 are determined by use of Eqs. 2.109, 2.110, and 2.111.

$$c_1 = -2\ (c_s - c_b)/\sqrt{\pi}, \quad c_2 = (c_s - c_b) \tag{2.113}$$

Then, the solution is obtained as Eq. 2.114.

$$c' = f = (c_s - c_b)\left\{1 - erf\left(z/2\sqrt{Dt}\right)\right\} \tag{2.114}$$

The key point in this method is to find the dimensionless variable η. The finding of η is not derived by chance, but can be done from the dimensionless procedure given in Sect. 3.

(2) The method to solve the partial differential equation 2.37 by use of the Laplace transform.

The Laplace transform $\mathcal{L}[c'\ (z, t)] \equiv C\ (z, s)$ of Eq. 2.37 with respect to t is given as:

$$sC\ (z,\ s) - C\ (z,\ 0) = D\, d^2C\ (z,\ s)/d\, z^2 \tag{2.115}$$

From Eq. 2.40,

$$\mathcal{L}[c'\ (z,\ 0)] \equiv C\ (z,\ 0) = 0 \tag{2.116}$$

so that Eq. 2.115 is reduced to Eq. 2.117.

$$d^2C\ (z,\ s)/d\, z^2 - (s/D)\ C\ (z,\ s) = 0 \tag{2.117}$$

The solution of this second-order ordinary differential equation is given as Eq. 2.118.

$$C\ (z,\ s) = A\ exp\left\{\sqrt{s/D}\ \ z\right\} + B\ exp\left\{\sqrt{-(s/D)}\ \ z\right\} \tag{2.118}$$

The Laplace transform of the boundary conditions are given as:

$$C\ (0,\ s) = (c_s - c_b)/s \tag{2.119}$$

$$C\ (\infty,\ s) = 0 \tag{2.120}$$

By substituting Eq. 2.119 into Eq. 2.118, the following relation is obtained.

$$(c_s - c_b)/s = A + B \tag{2.121}$$

From Eqs. 2.118, 2.120, and 2.121, A and B are obtained as:

$$A = 0, \tag{2.122}$$

$$B = (c_s - c_b)/s \tag{2.123}$$

Then, C (z,s) is given as:

$$C(z, s) = \{(c_s - c_b)/s\} \exp\left\{-\sqrt{(s/D)}\ z\right\} = \{(c_s - c_b)/s\} \exp\left\{-(z/\sqrt{D})\sqrt{s}\right\} \tag{2.124}$$

By use of the inverse Laplace transform defined as $\mathcal{L}^{-1}\{(1/s)\, exp(-a\sqrt{s}) = erfc\,(a/2\sqrt{t})$, Eq. 2.41 is obtained as follows:

$$c\ (z,\ t) = (c_s - c_b)\, erfc\left\{(z/\sqrt{D})/(2\sqrt{t})\right\} = (c_s - c_b)\left\{1 - \text{erf}\ (z/2\sqrt{Dt})\right\} \tag{2.41}$$

Appendix B: Derivation of Eq. 2.54

Let us put $c = c_s + c_2$. Then, c_2 should be satisfied by the governing equation (2.125) under the boundary and initial conditions of Eqs. 2.126, 2.127, and 2.128, which are modified from Eqs. 2.51, 2.52, and 2.53 as follows:

$$\partial c_2/\partial t = D\ \partial^2 c_2/\partial z^2 \tag{2.125}$$

$$\partial c_2/\partial z\ |_{z=0} = 0 \tag{2.126}$$

$$c_2\ |_{z=l} = 0 \tag{2.127}$$

$$c_2\ |_{t=0} = c_b - c_S \tag{2.128}$$

Putting $c_2(z, t) = Z(z) \cdot T(t)$ and substituting it into Eq. 2.125, then the following equation is obtained.

$$(1/T)(dT/dt) = (D/Z)(d^2T/dz^2), \tag{2.129}$$

where $T(t)$ and $Z(z)$ can be obtained by solving the following equations, respectively.

$$dT/dt - DkT = 0 \tag{2.130}$$

$$d^2Z/dz^2 - kZ = 0 \tag{2.131}$$

From Eq. 2.130, $T = C\ exp\ (Dkt)$ is the solution.

On the other hand, the solution of $Z = az + b$ when $k = 0$ is obtained from (2.131). However, it is found that this solution is not suitable since $Z = 0$ is from $a = b = 0$ which are obtained from Eqs. 2.126 and 2.127. If we put $k = p^2$, where k is positive and p is a real number, the solution of $Z = A\ exp\ (pz) + B\ exp\ (-pz)$ is obtained. This solution is also not suitable since $A = B = 0$ is from Eqs. 2.126 and 2.127. Then, let k be negative, that is $k = -q^2$. The solution is obtained as $Z = A\ cos\ (qz) + B\ sin\ (qz)$. From (2.126), $dZ/dz = Bq = 0$ is so that $B = 0$ yields. Then, the solution is $Z = A\ cos\ (qz)$. From Eq. 2.127, the condition of $A\ cos\ (ql) = 0$ is obtained. That is, q should be satisfied in the following condition.

$$cos(ql) = 0 \tag{2.132}$$

That is, the condition which q should be satisfied is as follows:

$$q = (2m + 1)\,\pi/2l,\ (m \text{ is integers}) \tag{2.133}$$

Then, c_2 is obtained as Eq. 2.134.

$$c_2 = A_m\ exp\ (-Dq^2t) \cdot \cos\ (qz), \tag{2.134}$$

where A_m is a function of m.

As Eq. 2.134 holds for all of integers m, c_2 is given as:

$$c_2 = \sum_{m=0}^{\infty} A_m\ exp\left\{-D(2m+1)^2\pi^2 t/4l^2\right\} \cdot \cos\{(2m+1)\pi/2l\}\,z \tag{2.135}$$

Finally, the function of A_m is determined to satisfy the initial condition of Eq. 2.128.

$$c_2\ |_{t=0} = \sum_{m=0}^{\infty} A_m \cdot \cos\ \{(2m+1)\pi z/2l\} = c_b - c_s,\ |z|\ < l \tag{2.136}$$

As the cosine is the even function, a Fourier series transform is applied over $0 < z < l$. That is,

$$\begin{aligned} A_m &= (2/l) \int_0^l (c_b - c_s) \cdot \cos\{(2m+1)\pi z/2l\}\, dz \\ &= \{2\,(c_b - c_s)/l\} \int_0^l \cos\ \{(2m+1)\pi z/2l\}\, dz \\ &= \{4\,(c_b - c_s)/(2m+1)\pi\}\ (-1)^m \end{aligned} \tag{2.137}$$

Then, c_2 is obtained as Eq. 2.138.

$$c_2 = \sum_{m=0}^{\infty} \{4\ (c_b - c_s)/(2m+1)\pi\}\ (-1)^m$$
$$\cdot\, exp\left\{-D\ (2m+1)^2\pi^2 t/4l^2\right\} \times \cos\{(2m+1)\pi z/2l\} \qquad (2.138)$$

The final solution is $c = c_s + c_2$, so that

$$c = c_s + 4\{(c_b - c_s)/\pi\} \sum_{m=0}^{\infty} \{1/(2m+1)\}\ (-1)^m$$
$$\times\, exp\left\{-D\ (2m+1)^2\pi^2 t/4l^2\right\} \cdot \cos\{(2m+1)\pi z/2l\} \qquad (2.54)$$

Appendix C: Derivation of Eq. 2.57

By putting $c = c_2 + c_b$, Eq. 2.33 and Eqs.2.51, 2.52, and 2.53 become as Eq. 2.139 and Eqs. 2.140, 2.141, and 2.142, respectively.

$$\partial c_2/\partial t = D\ \partial^2 c_2/\partial z^2 \qquad (2.139)$$

$$\partial c_2/\partial z\ |_{z=0} = 0 \qquad (2.140)$$

$$c_2\ |_{z=\pm l} = c_s - c_b \qquad (2.141)$$

$$c_2\ |_{t=0} = 0 \qquad (2.142)$$

The Laplace transform $\mathcal{L}\,[c_2(z, t)] \equiv C_2(z, s)$ of Eq. 2.139 with respect to t is given as Eq. 2.143.

$$s\,C_2(z,\ s) - C_2(z,\ 0) = D\ d^2C_2(z,\ s)/d\ z^2 \qquad (2.143)$$

Also, the Laplace transforms of Eqs. 2.140, 2.141, and 2.142 are given as:

$$d\ C_2(0,\ s)/dz = 0 \qquad (2.144)$$

$$C_2(x,\ s)\ |_{z=\pm l} = (c_s - c_b)/s \qquad (2.145)$$

$$C_2(z,\ 0) = 0 \qquad (2.146)$$

The general solution of Eq. 2.143 is obtained as follows:

$$C_2(z,\ s) = A\ exp\left(-\sqrt{s/D}\ z\right) + B exp\left(\sqrt{s/D}\ z\right) \tag{2.147}$$

From the boundary conditions of Eqs. 2.144 and 2.145, $A = B$ and the following relation are obtained.

$$C_2(l,\ s) = A\left\{exp\left(-\sqrt{s/D}\ l\right) + exp\left(\sqrt{s/D}\ l\right)\right\} = (c_s - c_b)/s \tag{2.148}$$

Thus,

$$A = B = (c_s - c_b)\Big/\left[s\left\{exp\left(-\sqrt{s/D}\ l\right) + exp\left(\sqrt{s/D}\ l\right)\right\}\right] \tag{2.149}$$

Substituting Eq. 2.149 into Eq. 2.147 yields Eq. 2.150.

$$\begin{aligned} C_2(z,\ s) &= (c_s - c_b)\left\{\exp\left(-\sqrt{s/D}\ z\right) + \exp\left(\sqrt{s/D}\ z\right)\right\}\Big/ \\ &\quad \left[s\left\{\exp\left(-\sqrt{s/D}\ l\right) + exp\ \left(\sqrt{s/D}\ l\right)\right\}\right] \\ &= \{(c_s - c_b)/D\}\cdot\left[\cosh\left(\sqrt{s/D}\ z\right)/\left\{(s/D)\cosh\left(\sqrt{s/D}\ l\right)\right\}\right] \end{aligned} \tag{2.150}$$

The Taylor expansion of $(1 + e^x)^{-1}$ is given as follows:

$$(1 + e^x)^{-1} = 1 - e^x + e^{2x} - e^{3x} + \cdots + (-1)^n e^{nx}, \tag{2.151}$$

so that Eq. 2.150 can be written as:

$$\begin{aligned} C_2(z,\ s) &= \{(c_s - c_b)/s\}\cdot\left(e^{-\sqrt{s/D}\ z} + e^{\sqrt{s/D}\ z}\right)\cdot\sum_{n=0}^{\infty}(-1)^n e^{-(2n+1)\sqrt{s/D}\ l} \\ &= \{(c_s - c_b)/s\}\cdot\sum_{n=0}^{\infty}(-1)^n\ e^{-\sqrt{s/D}\ z}e^{-(2n+1)\sqrt{s/D}\ l} \\ &\quad + \{(c_s - c_b)/s\}\cdot\sum_{n=0}^{\infty}(-1)^n e^{\sqrt{s/D}\ z}e^{-(2n+1)\sqrt{s/D}\ l} \\ &= \{(c_s - c_b)/s\}\cdot\sum_{n=0}^{\infty}(-1)^n e^{-\sqrt{s/D}\{(2n+1)l+z\}} \\ &\quad + \{(c_s - c_b)/s\}\cdot\sum_{n=0}^{\infty}(-1)^n e^{-\sqrt{s/D}\{(2n+1)l-z\}} \end{aligned} \tag{2.152}$$

The inverse Laplace transformation gives the following relation as:

$$\mathcal{L}^{-1}\left\{(1/s)\cdot\exp\left(-a\sqrt{s}\right)\right\} = erfc\left(a/2\sqrt{t}\right) \tag{2.153}$$

By using the above relation, the inverse Laplace transformation of Eq. 2.152 is written as:

$$\begin{aligned} c_2 = (c_s - c_b)\sum_{n=0}^{\infty}(-1)^n\, erfc\left\{\frac{(2n+1)l+z}{2\sqrt{Dt}}\right\} \\ + (c_s - c_b)\sum_{n=0}^{\infty}(-1)^n erfc\left\{\frac{(2n+1)l-z}{2\sqrt{Dt}}\right\} \end{aligned} \tag{2.154}$$

Finally, the solution c is obtained from $c = c_2 + c_b$ as follows:

$$\begin{aligned} c = c_b + (c_s - c_b)\Bigg[\sum_{n=0}^{\infty}(-1)^n erfc\left\{\frac{(2n+1)l+z}{2\sqrt{Dt}}\right\} \\ + \sum_{n=0}^{\infty}(-1)^n erfc\left\{\frac{(2n+1)l-z}{2\sqrt{Dt}}\right\}\Bigg] \end{aligned} \tag{2.57}$$

Appendix D: Derivation of Eq. 2.65

By putting $\theta = c - c_s$, Eqs. 2.61, 2.62, 2.63, and 2.64 are written as follows:

$$\partial\theta/\partial t = D\left\{\partial^2\theta/\partial r^2 + (1/r)\cdot(\partial\theta/\partial r)\right\} \tag{2.155}$$

$$\partial\theta/\partial r = 0, \quad \text{at} \quad r = 0, \quad t \geq 0 \tag{2.156}$$

$$\theta = 0, \quad \text{at} \quad r = R, \quad t \geq 0 \tag{2.157}$$

$$\theta = c_b - c_s \quad \text{at} \quad t = 0, \quad 0 \leq r \leq R \tag{2.158}$$

We may attempt to find a solution of Eq. 2.155 by putting $\theta(r, t) = X(r)\cdot T(t)$, where the X and T are functions of r and t, respectively. Equations 2.159 and 2.160 are obtained as follows:

$$X'' + (1/r)X' - kX = 0 \tag{2.159}$$

$$T' - kDT = 0, \tag{2.160}$$

where k is constant.

From Eqs. 2.156 and 2.157, the following boundary conditions are obtained.

$$X'(0) = 0 \tag{2.161}$$

$$X(R) = 0 \tag{2.162}$$

In the case of $k = 0$, the solution in the ordinary differential equation of Eq. 2.159 is $X = C_1 In(r) + C_2$. In this case, from the above boundary conditions, $C_1 = C_2 = 0$ is obtained. Then, the solution is $X = 0$. That is, it is understood that $k = 0$ does not provide a suitable solution. In the case of $k > 0$, the solution of Eq. 2.159 is $X = C_3 I_0(\sqrt{k}r) + C_4 K_0(\sqrt{k}r)$, where I_0 and K_0 are the modified Bessel functions of the first and second kinds of order zero, respectively. From the boundary condition of Eq. 2.161 with a finite value and the conditions of $I_0'(0) = I_1(0) = 0$ and $K_0'(0) = -K_1 = -\infty$, so that $C_4 = 0$ is obtained, and from the boundary conditions of Eq. 2.162 and the condition of $I\ (0)|_{r=0} \neq 0$, $C_3 = 0$ is. Thus, the condition of $k > 0$ is also an unsuitable choice since $X = 0$ is. Finally, in the case of $k < 0$, the solution of Eq. 2.159 is $X = C_5 J_0(\sqrt{-k}r) + C_6 Y_0(\sqrt{-k}r)$, where J_0 and Y_0 are Bessel functions of the first and second kinds of order zero, respectively. From the boundary condition of Eq. 2.161 and the conditions of $J_0'(0) = \mathit{finite}$ and $Y_0'(0) = \infty$, so that $C_6 = 0$ is obtained, and from the boundary condition of Eq. 2.162, $C_5 J_0(\sqrt{-k_n}R) = 0$ must be held, where k_n should be the n-th positive root of $J_0(\sqrt{-k_n}R) = 0$. Then, X is given as follows:

$$X = \sum_{n=1}^{\infty} J_0(\lambda_n r), \quad \lambda_n \equiv \sqrt{-k_n} \tag{2.163}$$

The other hand, from Eq. 2.160, the function of T is given as:

$$T = C_7 exp\left(-\lambda_n^2 Dt\right) \tag{2.164}$$

Then, $\theta(r, t)$ is obtained as Eq. 2.165.

$$\theta\ (r, t) = \sum_{n=1}^{\infty} C_n J_0(\lambda_n r) exp\left(-\lambda_n^2 Dt\right) \tag{2.165}$$

From the initial condition of Eq. 2.158, the following condition is obtained.

$$\theta\ (r,\ 0) = \sum_{n=1}^{\infty} C_n J_0(\lambda_n r) = c_b - c_s \tag{2.166}$$

In order to determine the values of C_n, $\int_0^R r J_0(\lambda_n r) dr$ is multiplied on both sides of Eq. 2.166 and integrated from 0 to R. By using the following characteristics of the orthogonal functions Eqs. 2.167, 2.168, and 2.169, the values of C_n are obtained as Eq. 2.170.

$$\int_0^R r J_0(\lambda_n r)\, J_0(\lambda_m r)\, dr = 0, \quad \mathrm{n} \neq \mathrm{m} \tag{2.167}$$

$$\int_0^R r\{J_0(\lambda_n r)\}^2 dr = (1/2) R^2 J_1^2(\lambda_n R) \tag{2.168}$$

$$\int_0^R r J_0(\lambda_n r)\, dr = R J_1(\lambda_n R)/\lambda_n, \tag{2.169}$$

where $J_1(x)$ is the Bessel function of the first order,

$$C_n = (c_b - c_s) \int_0^R r J_0(\lambda_n r)\, dr \bigg/ \int_0^R r\{J_0(\lambda_n r)\}^2 dr = 2(c_b - c_s)/R\lambda_n J_1(\lambda_n R) \tag{2.170}$$

Substituting Eq. 2.170 into Eq. 2.165 and using the relation of $\theta = c - c_s$ yields Eq. 2.65.

$$c = \{2\ (c_b - c_s)/R\} \sum_{n=1}^{\infty} \{J_0(\lambda_n r)/\lambda_n J_1(\lambda_n R)\} \cdot exp\left(-D\lambda_n^2 t\right) + c_s \tag{2.65}$$

Appendix E: Derivation of Eq. 2.72

By setting $\eta(z, y) \equiv (c - c_s)/(c_b - c_s), y \equiv r/R$, and $z \equiv Dt/R^2$ to reduce to non-dimensional form, Eq. 2.68 is expressed as Eq. 2.171.

$$\partial\eta/\partial z = \partial^2\eta/\partial y^2 + (2/y) \cdot (\partial\eta/\partial y), \quad (0 < z < \infty, \quad 0 < y < 1) \tag{2.171}$$

From the initial and boundary conditions, $\eta\ (0, y) = 1$, $\eta\ (z, 1) = 0$, $\eta\ (z, 0) = \partial\eta/\partial y|_{y=0} = 0$ are obtained. The right-hand side of Eq. 2.171 is modified by multiplying y as follows:

$$\begin{aligned} y(\partial^2\eta/\partial y^2) + 2 \cdot (\partial\eta/\partial y) &= y(\partial^2\eta/\partial y^2) + (\partial\eta/\partial y) + (\partial\eta/\partial y) \\ &= \partial\{y(\partial\eta/\partial y) + \eta\}/\partial y = \partial^2(\eta y)/\partial y^2 \end{aligned} \tag{2.172}$$

And multiplying y on the left-hand side of Eq. 2.171 yields Eq. 2.173.

$$y(\partial\eta/\partial z) = \partial(\eta y)/\partial z \tag{2.173}$$

That is, by introducing $\varphi(z, y) \equiv \eta y$, Eq. 2.171 can be expressed as Eq. 2.174.

$$\partial\varphi/\partial z = \partial^2\varphi/\partial y^2 \tag{2.174}$$

Then, the initial and boundary conditions become as follows:

$$\varphi(0, y) = y, \quad \varphi(z, 1) = 0, \quad \varphi(z, 0) = 0$$

We may attempt to find a solution of Eq. 2.174 by putting $\varphi = Z(z) \cdot Y(y)$, where Z and Y are functions of z and y, respectively. Substitution of $\varphi = Z \cdot Y$ in Eq. 2.174 yields Eq. 2.175.

$$(1/Z) \cdot (dZ/dz) = (1/Y) \cdot (d^2Y/dy^2) = -\alpha^2, \quad (\alpha \text{ is a real number}) \tag{2.175}$$

Then, we get the two ordinary differential equations as follows:

$$dZ/dz + \alpha^2 Z = 0 \tag{2.176}$$

$$d^2Y/dy^2 + \alpha^2 Y = 0 \tag{2.177}$$

The corresponding general solutions are shown as follows:

$$Z = C_1 \, exp\,(-\alpha^2 z) \tag{2.178}$$

$$Y = C_2 \cos\,(\alpha y) + C_3 sin\,(\alpha y) \tag{2.179}$$

From the boundary conditions, the following conditions are obtained.

$$C_2 = 0, \qquad \text{from } \varphi(z, 0) = 0 \tag{2.180}$$

$$sin\,\alpha = 0, \quad \text{from } \varphi(z, 1) = 0 \tag{2.181}$$

From Eq. 2.181, α should be given as:

$$\alpha = n\pi, \qquad (n \text{ expresses integer numbers}) \tag{2.182}$$

Thus, the general solution of φ is given as:

$$\varphi = \sum_{n=1}^{\infty} C_n exp\,(-n^2\pi^2 z) \cdot sin(n\pi y) \tag{2.183}$$

C_n is determined to satisfy the condition of $\varphi(0, y) = y$ by using the Fourier sine transform.

$$\varphi = \sum_{n=1}^{\infty} C_n sin\ (n\pi y) = y \tag{2.184}$$

$$\text{Thus,}\quad C_n = 2\int_0^1 \xi si\,n(n\pi\xi)\ d\xi = (2/n\pi)\cdot(-1)^{n+1} \tag{2.185}$$

Then, the solution of φ is obtained as:

$$\varphi = (2/\pi)\sum_{n=1}^{\infty}\left\{(-1)^{n+1}/n\right\}\cdot exp\ (-n^2\pi^2 z)\cdot sin\ (n\pi y) \tag{2.186}$$

Finally $\varphi(z, y)$ is reduced to $c\ (r, t)$ and then Eq. 2.72 is obtained.

$$c = c_s + \{2R\ (c_b - c_s)/\pi r\}\sum_{n=1}^{\infty}\left\{(-1)^{n+1}/n\right\} exp\,(-n^2\pi^2 Dt/R^2) sin(n\pi r/R) \tag{2.72}$$

References

1. M. Zahn, *Electromagnetic Field Theory* (Wiley, New York, 1979), pp. 3, 5, 6
2. R.B. Bird, W.E. Stewart, E.N. Lightfoot, *Transport Phenomena*, 2nd edn. (Wiley, New York, 2002), p. 336
3. *i.bi.dem*, p. 847
4. *i.bi.dem*, p. 152
5. R. Moreau, *Magnetohydrodynamics* (Kluwer Academic, Dordrecht/Boston, 1990), p. 290
6. P.A. Davidson, *An Introduction to Magnetohydrodynamics* (Cambridge University Press, Cambridge/New York, 2001), pp. 128, 249
7. K. Takatani, *The 5th International Symposium on Electromagnetic Processing of Materials* (ISIJ, Tokyo, 2006), p. 176

Chapter 3
Magnetohydrodynamics in Materials Processing

Magnetohydrodynamics was established by Alfvén in 1942. When we solve problems concerning a flow of highly conductive fluids such as molten metals or salts under an electric and/or magnetic fields, an electromagnetic force term induced by electric and magnetic fields appears in a force term in the equation of motion. Therefore, it is necessary to know the directions and magnitudes of the existing electric and magnetic fields. On the other hand, to obtain those parameters in electric and magnetic fields in the highly conductive fluid by solving Maxwell's equation, we have to know the fluid velocity in a generalized Ohm's law. In this way, magnetohydrodynamics describes problems where the electric and magnetic fields couple with the velocity field. Magnetohydrodynamics covers not only plasma physics, geophysics and astrophysics but also EPM (Electromagnetic Processing of Materials), which is the main subject of this book. The mutual interaction between the electric and magnetic fields and the velocity field depends on the problems we have to solve. In most cases of EPM, it is generally to be considered that the electric and magnetic fields influence the velocity field, while the velocity field scarcely influences the electric and magnetic fields. Therefore, we first solve the equation of electric and magnetic fields without considering the influence of the velocity field, and then get the velocity field by substituting the obtained electric and magnetic fields into the driving force term in the equation of motion. That is, the coupling will be one-sided in many cases.

In this chapter, the subjects that are indispensable for explaining EPM are selected from the view-point of MHD (magnetohydrodynamics) on incompressible Newtonian fluids with high electric conductivity such as molten metals.

S. Asai, *Electromagnetic Processing of Materials*, Fluid Mechanics and Its Applications 99, DOI 10.1007/978-94-007-2645-1_3,
© Springer Science+Business Media B.V. 2012

1 Fundamental Equation

1.1 Transport Equation of Vorticity

Let us consider the physical meaning of a vorticity before deriving the transport equation. Equation 3.1 is the definition of a vorticity, which is a vector quantity.

$$\boldsymbol{\omega} = \nabla \times \boldsymbol{v}, \quad \boldsymbol{\omega} = (\omega_r, \omega_\theta, \omega_z) \tag{3.1}$$

Now, as shown in Fig. 3.1, when a flow is circulating around the origin of a coordinate with an angular velocity λ, the components of the circulating velocity are expressed by $v_r = 0$, $v_\theta = \lambda r$, $v_z = 0$. Meanwhile, we can get the following results by substituting the components of the circulating velocity into Eq. 3.1 described by the vector identity

$$\omega_r = \frac{1}{r}\frac{\partial v_z}{\partial \theta} - \frac{\partial v_\theta}{\partial z} = 0, \quad \omega_\theta = \frac{\partial v_r}{\partial z} - \frac{\partial v_z}{\partial r} = 0, \quad \omega_z = \frac{1}{r}\left\{\frac{\partial (r v_\theta)}{\partial r} - \frac{\partial v_r}{\partial \theta}\right\} = 2\lambda$$

From these results, it can be seen that the vorticity vector has a vertical component to the circulation plane (z-direction). Its magnitude is two times the angular velocity λ.

Let us consider the circulating flow in the sequence of A → B → C → D → A around a minute square on an x-y plane as shown in Fig. 3.2. A sum of (average velocity) × (length of motion) in the flow along each path is called circulation. In mathematics it is defined as $C = \oint (v_x dx + v_y dy + v_z dz)$. In the case of Fig. 3.2, the circulation is calculated as follows:

$$\begin{aligned} C = & \left(v_x + \frac{1}{2}\frac{\partial v_x}{\partial x}dx\right)dx + \left(v_y + \frac{\partial v_y}{\partial x}dx + \frac{1}{2}\frac{\partial v_y}{\partial y}dy\right)dy \\ & - \left(v_x + \frac{\partial v_x}{\partial y}dy + \frac{1}{2}\frac{\partial v_x}{\partial x}dx\right)dx - \left(v_y + \frac{1}{2}\frac{\partial v_y}{\partial y}dy\right)dy = \left(\frac{\partial v_y}{\partial x} - \frac{\partial v_x}{\partial y}\right)dxdy \end{aligned}$$

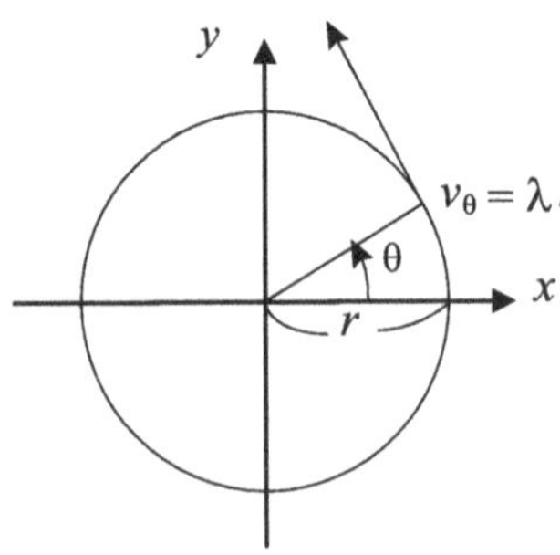

Fig. 3.1 A circulating flow with an angular velocity λ

Fig. 3.2 Circulation around a minute square ABCD

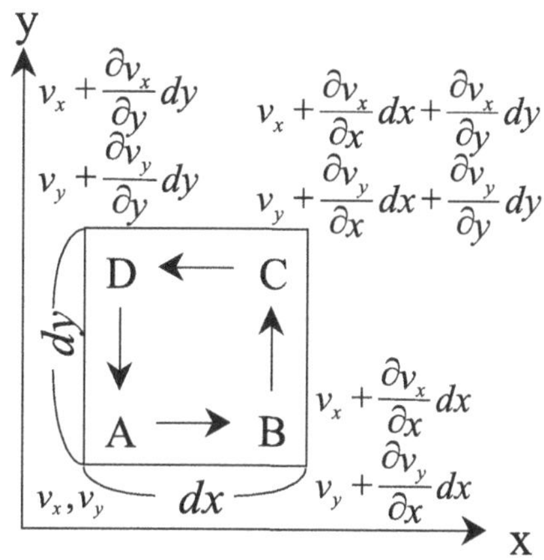

By dividing C with the area of *dxdy*,

$$C/dxdy = (\partial v_y/\partial x) - (\partial v_x/\partial y) \equiv \omega_z$$

is obtained. That is, the circulation corresponds to z-component ω_z of the vorticity expressed on the rectangular coordinate.

$$\omega_x = \frac{\partial v_z}{\partial y} - \frac{\partial v_y}{\partial z}, \quad \omega_y = \frac{\partial v_x}{\partial z} - \frac{\partial v_z}{\partial x}, \quad \omega_z = \frac{\partial v_y}{\partial x} - \frac{\partial v_x}{\partial y}$$

Hence, the vorticity is equal to the quotient that is obtained by dividing the circulation (velocity × distance) of a fluid element by the area surrounded by the fluid element motion.

The equation of motion for an incompressible Newtonian fluid, which is called the Navier-Stokes equation, is given from Eq. 2.32.

$$\rho\{\partial \boldsymbol{v}/\partial t + (\boldsymbol{v} \cdot \nabla)\boldsymbol{v}\} = -\nabla p + \mu \nabla^2 \boldsymbol{v} + \boldsymbol{f}, \tag{3.2}$$

where $\boldsymbol{f} = (f_x, f_y, f_z)$ is a term of body force acting on a unit volume of fluid.

Equation 3.3 can be derived by substituting an electromagnetic force $\boldsymbol{J} \times \boldsymbol{B}$ in the force term of Eq. 3.2 and then applying the vector operator ($\nabla \times$) to both sides. This vector operation is called rotation. In mathematics the operator ($\nabla \times$) is imposed from the left-hand side of the equation.

$$\underbrace{\partial \boldsymbol{\omega}/\partial t}_{\text{unsteady state term}} = \underbrace{\nabla \times (\boldsymbol{v} \times \boldsymbol{\omega})}_{\text{convection term}} + \underbrace{\nu \nabla^2 \boldsymbol{\omega}}_{\text{diffusion term}} + \underbrace{\frac{1}{\rho} \nabla \times (\boldsymbol{J} \times \mathbf{B})}_{\text{generation term}}, \quad \nu \equiv \mu/\rho \tag{3.3}$$

Problem 3.1. Derive Eq. 3.3 by using the following vector identity.

$$\begin{aligned}\nabla \times (\boldsymbol{v} \cdot \nabla)\boldsymbol{v} &= \nabla \times \{\nabla(\boldsymbol{v} \cdot \boldsymbol{v})/2 - \boldsymbol{v} \times (\nabla \times \boldsymbol{v})\} \\ &= \nabla \times \nabla (\boldsymbol{v} \cdot \boldsymbol{v})/2 - \nabla \times (\boldsymbol{v} \times \boldsymbol{\omega})\end{aligned}$$

1.2 The Equation of Energy

An equation of energy that takes into account a heat source term due to Joule heating can be derived from Eq. 2.18.

$$c_p \rho\{\partial T/\partial t + (\boldsymbol{v} \cdot \nabla)T\} = \lambda \nabla^2 T + J^2/\sigma, \tag{3.4}$$

where σ(S/m) is the electric conductivity and $\lambda\{\mathrm{J}/(\mathrm{m\,s\,K})\}$ the thermal conductivity.

1.3 The Diffusion Equation of the Magnetic Field

Under MHD approximation, Faraday's law is expressed as Eq. 3.5.

$$\nabla \times \boldsymbol{E} = -\partial \boldsymbol{B}/\partial t \quad \text{(Faraday's law)} \tag{3.5}$$

Equation 3.5 denotes that the gradient of an electric field $\boldsymbol{E}$ is induced by change of a magnetic field with time (the right-hand side indicates the cause; the left-hand side indicates the result).

Ampere's law is written as Eq. 3.6.

$$\nabla \times \boldsymbol{H} = \boldsymbol{J} \quad \text{(Ampere's law)} \tag{3.6}$$

Equation 3.6 shows that the gradient of the magnetic field $\boldsymbol{H}$ is induced by an electric current $\boldsymbol{J}$.

Gauss' law is given as Eq. 3.7.

$$\nabla \cdot \boldsymbol{B} = 0 \quad \text{(Gauss' law)} \tag{3.7}$$

Equation 3.7 expresses the continuity of a magnetic field and says that the quantity of the magnetic fluxes that enters in any areas is equal to that which goes out of the areas.

The other Gauss' law is shown in Eq. 3.8.

$$\nabla \cdot \boldsymbol{D} = \rho_f \quad \text{(Gauss' law),} \tag{3.8}$$

where $\boldsymbol{D}$ is the displacement field vector and given as $\boldsymbol{D} = \varepsilon\boldsymbol{E}$, and ρ_f and ε are called the net charge density and the dielectric constant, respectively.

Applying the operator of the divergence ($\nabla\cdot$) to both sides of Eq. 3.6 and using the vector identity of $\nabla\cdot(\nabla\times\boldsymbol{H}) = 0$ yields Eq. 3.9, which expresses the continuity of an electric current $\boldsymbol{J}$.

$$\nabla\cdot\boldsymbol{J} = 0 \tag{3.9}$$

The constitutional equation of a magnetic field can be written as Eq. 3.10,

$$\boldsymbol{B} = \mu_m\boldsymbol{H}, \tag{3.10}$$

where $\boldsymbol{B}$ is the magnetic flux density (T), $\boldsymbol{H}$ the magnetic field intensity (A/m) and μ_m the magnetic permeability (H/m).

As the motion of a conductive media in a magnetic field induces an electric current, Ohm's law is extended to the following equation.

$$\boldsymbol{J} = \sigma(\boldsymbol{E} + \boldsymbol{v}\times\boldsymbol{B}) \tag{3.11}$$

From Eqs. 3.5, 3.6, 3.7, 3.8, 3.9, 3.10, 3.11, the diffusion equation of a magnetic field can be derived as Eq. 3.12.

$$\underbrace{\partial\boldsymbol{B}/\partial t}_{\text{unsteady state term}} = \underbrace{\nabla\times(\boldsymbol{v}\times\boldsymbol{B})}_{\text{convection term}} + \underbrace{\nu_m\nabla^2\boldsymbol{B}}_{\text{diffusion term}}, \tag{3.12}$$

where ν_m ($\equiv 1/\sigma\mu_m$) is the magnetic diffusivity, which has the same dimension with that of solutes (m^2/s).

By comparing Eq. 3.3 with Eq. 3.12, it can be found that the two equations have the same formulation, besides no generation term in Eq. 3.12. The magnetic Prandtl number of a dimensionless number is defined as the ratio of the kinematic viscosity ν in Eq. 3.3 to ν_m in Eq. 3.12.

$$Pr_m = \nu/\nu_m \tag{3.13}$$

This number indicates the ratio of the thickness of both boundary layers in a laminar flow. In a liquid metal, Pr_m is much smaller than 1. This means that the thickness of the magnetic field boundary layer is much larger than that of the velocity one.

In a non-electrical conductive medium, Eq. 3.14 can be derived from Eq. 3.6 due to $\boldsymbol{J} = 0$.

$$\nabla\times\boldsymbol{H} = 0 \tag{3.14}$$

Equation 3.14 indicates that a magnetic scalar potential ϕ exists, which is defined by Eq. 3.15.

$$\boldsymbol{B} = -\nabla\phi \tag{3.15}$$

From Eqs. 3.7 and 3.15, Eq. 3.16 is derived.

$$\nabla^2 \phi = 0 \tag{3.16}$$

The solution ϕ can be obtained by solving the Laplace equation of Eq. 3.16 under arbitrary boundary conditions. Once a solution ϕ has been obtained, the distribution of the magnetic field can be obtained by substituting ϕ into Eq. 3.15. The solution ϕ of the Laplace equation is called a harmonic function, where the mathematical formulation can be found in mathematical books.

Problem 3.2. Derive Eq. 3.9 from Eq. 3.6.

Problem 3.3. Derive Eq. 3.12 from Eqs. 3.5, 3.6, 3.7, 3.8, 3.9, 3.10, and 3.11.

Problem 3.4. Explain why the scalar potential of a magnetic field ϕ exists when $\nabla \times \boldsymbol{H} = \boldsymbol{0}$

2 Electromagnetic Force and Maxwell Stress Tensor

Lorentz force $\boldsymbol{f}$ is given by Eq. 3.17

$$\boldsymbol{f} = \boldsymbol{J} \times \boldsymbol{B} \tag{3.17}$$

Substituting Eqs. 3.6 and 3.10 into Eq. 3.17 and using the vector identity yield Eq. 3.18.

$$\begin{aligned} \boldsymbol{f} &= \boldsymbol{J} \times \boldsymbol{B} = (1/\mu_m)(\nabla \times \boldsymbol{B}) \times \boldsymbol{B} \\ &= \underbrace{(1/\mu_m)(\boldsymbol{B} \cdot \nabla)\boldsymbol{B}}_{\text{rotational component}} - \underbrace{\nabla(B^2/2\mu_m)}_{\text{irrotational component}} \end{aligned} \tag{3.18}$$

As understood from Eq. 3.3 expressed by use of vorticity, the fluid motion is caused not by the force acting on the fluid $\boldsymbol{f}$, but by the gradient, namely rotation of the force $\boldsymbol{f}$. In order to understand this mathematical derivation in a physical image, let us consider water in a cup as an example. The water receives gravity force, but it does not make a motion in the cup. This is due to the gravity force uniformly acting on the water. That is, the source of the vorticity is zero, which is deduced from the result that the rotation of the gravity force $\rho \boldsymbol{g}$ is zero. A field in which the rotation is zero, such as the gravity field, is called a potential field. On the other hand, a temperature distribution exists in hot water in a cup so that the rotation of the gravity force $\rho \boldsymbol{g}$ is not zero due to the unevenness of density ρ in it. Thus a motion takes place in the fluid, the so-called natural convection. In this case, the temperature distribution in the fluid breaks the potential field.

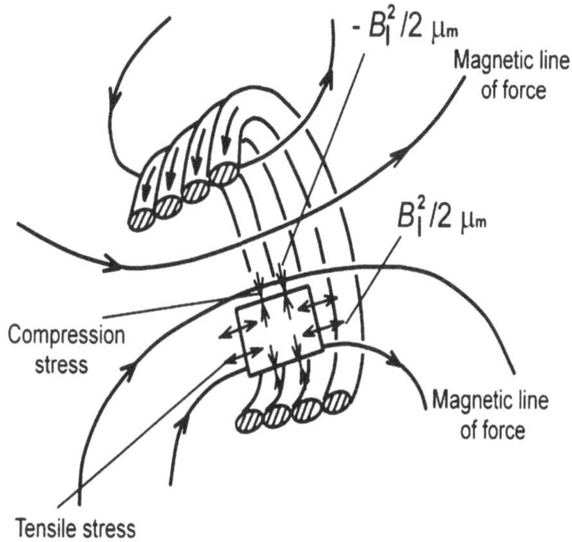

Fig. 3.3 Tensile and compression stresses caused by a magnetic field

By taking the rotation of the f, we can find out which component of the Lorentz force contributes to a fluid motion.

$$\nabla \times \boldsymbol{f} = \nabla \times (1/\mu_m)(\boldsymbol{B} \cdot \nabla)\boldsymbol{B} - \nabla \times \nabla(B^2/2\mu_m) = (1/\mu_m)\nabla \times (\boldsymbol{B} \cdot \nabla)\boldsymbol{B} \tag{3.19}$$

From Eq. 3.19, the origin of the rotation and irrotation components will be clear.

A magnetic field imposed on a fluid exerts a stress on the fluid. This stress is called Maxwell stress and its stress tensor σ_{ij} is defined as follows:

$$\sigma_{ij} = (1/\mu_m)\{B_i B_j - (\delta_{ij}|\boldsymbol{B}|^2/2)\} \tag{3.20}$$

In Eq. 3.20, δ_{ij} is called the Kronecker delta. When i equals j, it takes the value of 1 and when i does not equal j, it takes the value of 0. Supposing that $\boldsymbol{B} = (B_l, 0, 0)$, from Eq. 3.20 the components of σ_{11}, σ_{22} and σ_{33} are $B_1^2/2\mu_m$, $-B_1^2/2\mu_m$, and $-B_1^2/2\mu_m$, respectively. Thus, it is understood that a tensile stress acts in the direction of the imposed magnetic field, and a compressive stress works in the direction perpendicular to the tensile stress as shown in Fig. 3.3.

3 The Effect of a Static Magnetic Field (Direct Magnetic Field) on a Flow

A magnetic field is imposed on an electrically conductive fluid flowing in a rectangular duct from its outside. Under this condition, let us see what relations among an electric, magnetic and velocity field hold by use of the theoretical formula

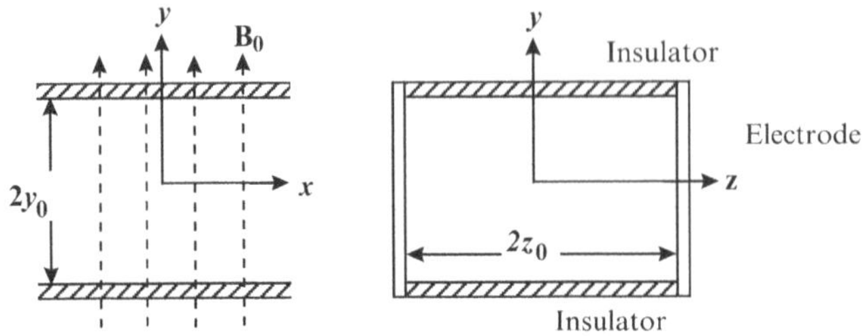

Fig. 3.4 Configuration showing the flow of a one-dimensional incompressible and electrically conductive fluid

derived in the previous sections. The system given here is shown in Fig. 3.4. An incompressible electrically conductive fluid (such as a liquid metal) flows in the x-direction through a duct with width $2z_0$ in the z-direction and thickness $2y_0$ in the y-direction. The magnetic field is imposed in the y-direction uniformly from outside. This is the problem of a one-dimensional incompressible flow under a magnetic field, which is called the Hartmann problem. Under the assumption of $z_0 \gg y_0$, all variables except p can be reduced to a function of only y.

The formulas of the continuation of the magnetic field and flow velocity are given in Eqs. 3.7 and 3.21.

$$\nabla \cdot \boldsymbol{B} = 0 \tag{3.7}$$

$$\nabla \cdot \boldsymbol{v} = 0 \tag{3.21}$$

Since $\boldsymbol{B} = (B_x, B_y, B_z)$ and $\boldsymbol{v} = (v_x, v_y, v_z)$ are functions of only y, Eqs. 3.7 and 3.21 are reduced to $\partial B_y/\partial y = 0$ and $\partial v_y/\partial y = 0$, i.e. $B_y = \text{const.}$ and $v_y = \text{const.}$ Then, Eqs. 3.22 and 3.23 are obtained.

$$B_y = B_0, \tag{3.22}$$

$$v_y = 0 \tag{3.23}$$

In this problem, as the magnetic field is not a function of time, i.e., $\partial \boldsymbol{B}/\partial t = 0$, Eq. 3.24 is obtained from Eq. 3.5.

$$\nabla \times \boldsymbol{E} = \left(\frac{\partial E_z}{\partial y} - \frac{\partial E_y}{\partial z}, \frac{\partial E_x}{\partial z} - \frac{\partial E_z}{\partial x}, \frac{\partial E_y}{\partial x} - \frac{\partial E_x}{\partial y} \right) = 0 \tag{3.24}$$

Since $\boldsymbol{E}$ is the function of only y, Eq. 3.24 is reduced to $\partial E_z/\partial y = 0$ and $\partial E_x/\partial y = 0$. Then, the following expressions are obtained.

$$E_z = \text{const.}, \tag{3.25}$$

$$E_x = \text{const.} \tag{3.26}$$

Next, $\nabla \times \boldsymbol{B} = \mu_m \boldsymbol{J}$ is formulated as follows:

$$\left(\frac{\partial B_z}{\partial y} - \frac{\partial B_y}{\partial z}, \frac{\partial B_x}{\partial z} - \frac{\partial B_z}{\partial x}, \frac{\partial B_y}{\partial x} - \frac{\partial B_x}{\partial y}\right) = \mu_m (J_x, J_y, J_z) \tag{3.27}$$

Since $\boldsymbol{B}$ is the function of only y, Eq. 3.28 is obtained.

$$\partial B_z/\partial y = \mu_m J_x, \quad 0 = \mu_m J_y, \quad -\partial B_x/\partial y = \mu_m J_z \tag{3.28}$$

The electromagnetic body force $\boldsymbol{J} \times \boldsymbol{B}$ can be expressed by Eq. 3.29 based on Eqs. 3.22 and 3.28.

$$\boldsymbol{J} \times \boldsymbol{B} = \begin{vmatrix} \boldsymbol{i}_x & \boldsymbol{i}_y & \boldsymbol{i}_z \\ J_x & 0 & J_z \\ B_x & B_0 & B_z \end{vmatrix} = \{-B_0 J_z \boldsymbol{i}_x, (J_z B_x - J_x B_z)\boldsymbol{i}_y, J_x B_0 \boldsymbol{i}_z\} \tag{3.29}$$

Then, the Navier-Stokes equation of Eq. 3.2 is formulated in each component as follows:

$$\text{x} - \text{component}: 0 = -\partial p/\partial x + \mu\, \partial^2 v_x/\partial y^2 - B_0\, J_z \tag{3.30}$$

$$\text{y} - \text{component}: 0 = -\partial p/\partial y + J_z\, B_x - J_x\, B_z \tag{3.31}$$

$$\text{z} - \text{component}: 0 = -\partial p/\partial z + \mu \partial^2 v_z/\partial y^2 + J_x\, B_0 \tag{3.32}$$

$\boldsymbol{v} \times \boldsymbol{B}$ appearing in Ohm's law $\boldsymbol{J} = \sigma(\boldsymbol{E} + \boldsymbol{v} \times \boldsymbol{B})$ is described as Eq. 3.33.

$$\boldsymbol{v} \times \boldsymbol{B} = \begin{vmatrix} \boldsymbol{i}_x & \boldsymbol{i}_y & \boldsymbol{i}_z \\ v_x & 0 & v_z \\ B_x & B_0 & B_z \end{vmatrix} = \{-B_0 v_z \boldsymbol{i}_x, (v_z B_x - v_x B_z)\boldsymbol{i}_y, v_x B_0 \boldsymbol{i}_z\} \tag{3.33}$$

Therefore, $\boldsymbol{J} = (J_x, J_y, J_z)$ is expressed as follows:

$$J_x = \sigma\,(E_x - B_0\, v_z) \tag{3.34}$$

$$J_y = \sigma\,(E_y + v_z\, B_x - v_x\, B_z) \tag{3.35}$$

$$J_z = \sigma\,(E_z + v_x\, B_0) \tag{3.36}$$

By substituting Eq. 3.36 into Eq. 3.30, Eq. 3.37 is obtained.

$$0 = -\partial p/\partial x + \mu\, \partial^2 v_x/\partial y^2 - \sigma\,(E_z + v_x\, B_0)\, B_0 \tag{3.37}$$

On the other hand, Eq. 3.38 is obtained when both sides of Eq. 3.31 are partially differentiated by x.

$$0 = \partial^2 p/\partial x \partial y + \partial (J_z\ B_x - J_x\ B_z)\ /\ \partial x \tag{3.38}$$

Since $(J_z\ B_x - J_x\ B_z)$ is the function of only y, Eq. 3.38 becomes Eq. 3.39.

$$\partial^2 p/\partial x \partial y = 0 \tag{3.39}$$

When this equation is integrated, it becomes Eq. 3.40.

$$p = c_1\, x + f(y), \qquad \partial p/\partial x = c_1 \tag{3.40}$$

As $\partial p/\partial x$ is a constant in Eq. 3.37 and E_z is also a constant (Ref. Eq. 3.25), Eq. 3.37 is found as an ordinary differential equation because v_x is also a function of only y.

$$d^2 v_x/dy^2 - \underset{\text{constant}}{\underline{(\sigma/\mu) B_0{}^2}}\, v_x = \underset{\text{constant}}{\underline{(1/\mu)\{(\partial p/\partial x) + \sigma\, B_0\, E_z\}}} \tag{3.41}$$

Solving Eq. 3.41 under the boundary conditions of $d v_x/dy = 0$ in $y = 0$ and $v_x = 0$ in $y = y_0$ yields Eq. 3.42.

$$v_x = \left(\frac{y_0}{Ha}\right)^2 \cdot \left(\frac{1}{\mu}\frac{\partial p}{\partial x} + \frac{Ha}{y_0}\sqrt{\frac{\sigma}{\mu}}\ E_z\right) \cdot \left\{\frac{\cosh\,(Ha y/y_0)}{\cosh Ha} - 1\right\} \tag{3.42}$$

Here, $Ha \equiv y_0\ B_0\ \sqrt{\frac{\sigma}{\mu}}$ is a dimensionless number called a Hartmann number. In Eq. 3.42, v_x is formulated by the product of $(\frac{y_0}{Ha})^2 \cdot (\frac{1}{\mu}\frac{\partial p}{\partial x} + \frac{Ha}{y_0}\sqrt{\frac{\sigma}{\mu}}\ E_z)$ and $\{\frac{\cosh\,(Ha\,y\,/\,y_0)}{\cosh\,Ha} - 1\}$, which express the amplitude of v_x and velocity distribution in y-direction, respectively.

Furthermore, an average flow velocity $\bar{v}_x$ over a cross-sectional area of the duct is obtained as Eq. 3.43 from Eq. 3.42.

$$\bar{v}_x = \frac{1}{2y_0}\int_{-y_0}^{y_0} v_x\ dy = \left(\frac{y_0}{Ha}\right)^2 \left(\frac{1}{\mu}\frac{\partial p}{\partial x} + \frac{Ha}{y_0}\sqrt{\frac{\sigma}{\mu}} E_z\right)\left(\frac{\tanh\,Ha}{Ha} - 1\right) \tag{3.43}$$

Current density is obtained from Eqs. 3.36 and 3.42 as Eq. 3.44.

$$J_z = \sigma\, E_z \frac{\cosh\,(Ha\,y/y_0)}{\cosh\,Ha} + \frac{y_0}{Ha}\sqrt{\frac{\sigma}{\mu}}\frac{\partial p}{\partial x}\left\{\frac{\cosh(Ha\,y/y_0)}{\cosh Ha} - 1\right\} \tag{3.44}$$

Therefore, a total current per unit length in the x-direction becomes Eq. 3.45.

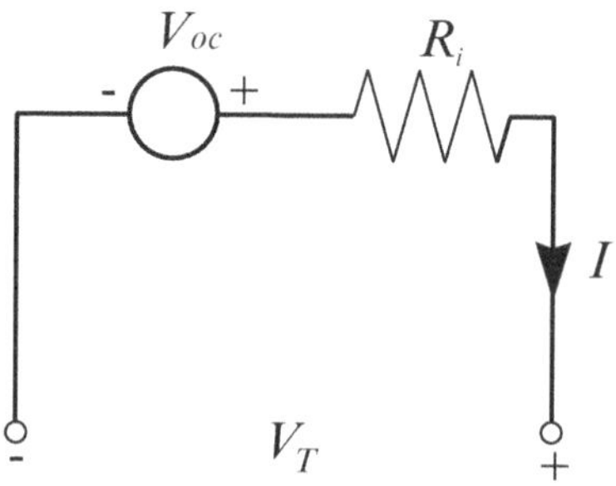

Fig. 3.5 An equivalent circuit of the Hartmann problem [1]

$$I_z = \int_{-y_0}^{y_0} J_z dy = 2\sigma y_0 E_z \frac{\tanh Ha}{Ha} + \frac{2y_0{}^2}{Ha}\sqrt{\frac{\sigma}{\mu}}\frac{\partial p}{\partial x}\left(\frac{\tanh Ha}{Ha} - 1\right) \quad (3.45)$$

This equation gives the relation between an electric field E_z and the current I_z. When Ha is asymptotic to zero, I_z is given by Eq. 3.46.

$$\lim_{\mathrm{Ha}\to\mathrm{o}} I_z = 2\sigma y_0 E_z \quad (3.46)$$

This result tells that I_z is independent of the pressure gradient $\partial p/\partial x$. Since the voltage between the electrodes is calculated as follows:

$$V_T = -\int_{-z_0}^{z_0} E_z dz = -2z_0 E_z \quad (3.47)$$

Equation 3.48 is obtained from Eqs. 3.45 and 3.47.

$$V_T = -\frac{z_0 Ha}{\sigma y_0 \tanh Ha} I_z + \frac{2y_0 z_0}{Ha\sqrt{\sigma\mu}}\frac{\partial p}{\partial x}\left(1 - \frac{Ha}{\tanh Ha}\right) \quad (3.48)$$

Equation 3.48 which describes the relation between the electric current and voltage can be expressed by an electric circuit, which is called an equivalent circuit. The equivalent circuit of a Hartmann problem expressed by Eq. 3.48 is shown in Fig. 3.5. Since the current is zero ($I_z = 0$) in an open circuit, the voltage is

$$V_{oc} = \frac{2y_0 z_0}{Ha\sqrt{\sigma\mu}}\frac{\partial p}{\partial x}\left(1 - \frac{Ha}{\tanh Ha}\right) \quad (3.49)$$

On the other hand, when it is a short-circuit between electrodes, $V_T = 0$ is valid. Then, the current is obtained from Eq. 3.48 as Eq. 3.50.

$$I_{sc} = \frac{2y_0{}^2 \times 1}{Ha}\sqrt{\frac{\sigma}{\mu}}\frac{\partial p}{\partial x}\left(\frac{\tanh Ha}{Ha} - 1\right) \quad (3.50)$$

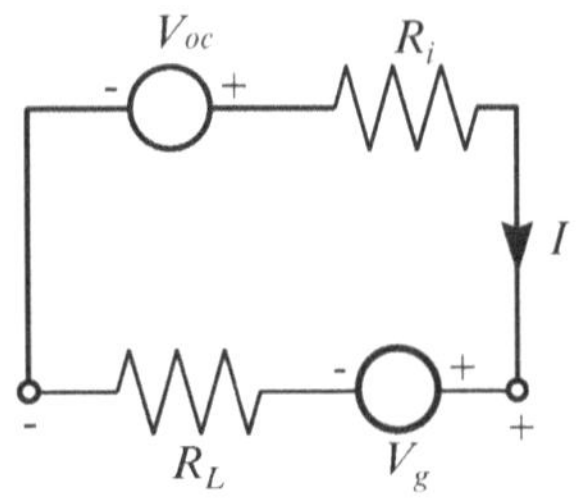

Fig. 3.6 The total electric circuits of the Hartmann problem [1]

The 1 appearing in $\times 1$ in the above equation means a unit width in the x-direction. An electric inner resistance R_i is obtained as follow:

$$R_i = V_{oc}/I_{sc} = \{z_0/(1 \times y_0)\sigma\}\,(Ha/\tanh\ Ha) \tag{3.51}$$

R_i is expressed as $R_i = z_0/(1 \times y_0)\sigma$ under the condition of $Ha \to 0$ and increases in proportion to Ha when $Ha > 3$.

On the other hand, when applying Kirchhoff's voltage law to the circuit shown in Fig. 3.6, we can obtain Eq. 3.52.

$$V_{oc} - I(R_i + R_L) - V_g = 0 \tag{3.52}$$

By substituting Eqs. 3.49 and 3.51 into V_{oc} and R_i in Eq. 3.52, respectively, the electric current in the circuit is obtained as Eq. 3.53.

$$I = \frac{-V_g + (2y_0z_0/Ha\sqrt{\sigma\mu})(\partial p/\partial x)\{1 - (Ha/\tanh Ha)\}}{R_L + (z_0/y_0\sigma)(Ha/\tanh Ha)} \tag{3.53}$$

When $Ha \to 0$, $\lim\limits_{Ha\to 0} I = -V_g/(R_i + R_L)$ is obtained, this means that the current is not induced in the flow when no magnetic field is imposed. Substituting I expressed by Eq. 3.53 into I_z in Eq. 3.48 and using the relation of $E_z = -V_T/2z_0$ yields E_z as follows:

$$E_z = \frac{(V_g/2\sigma y_0)(Ha/\tanh Ha) + (y_0R_L/Ha\sqrt{\sigma\mu})(\partial p/\partial x)\{1 - (Ha/\tanh Ha)\}}{-\{R_L + (z_0/y_0\sigma)(Ha/\tanh Ha)\}} \tag{3.54}$$

When there is no electric generator in the outside of the duct $V_g = 0$ and further $R_L = 0$, $E_z = 0$ holds. When $R_L \to \infty$, E_z becomes as follows:

$$\lim_{R_L\to\infty} E_z = -(y_0/Ha\sqrt{\sigma\mu})(\partial p/\partial x)\{1 - (Ha/\tanh Ha)\} \tag{3.55}$$

By substituting Eq. 3.54 into Eq. 3.42, an integrated expression for the velocity taking into account the external circuit can be obtained as Eq. 3.56.

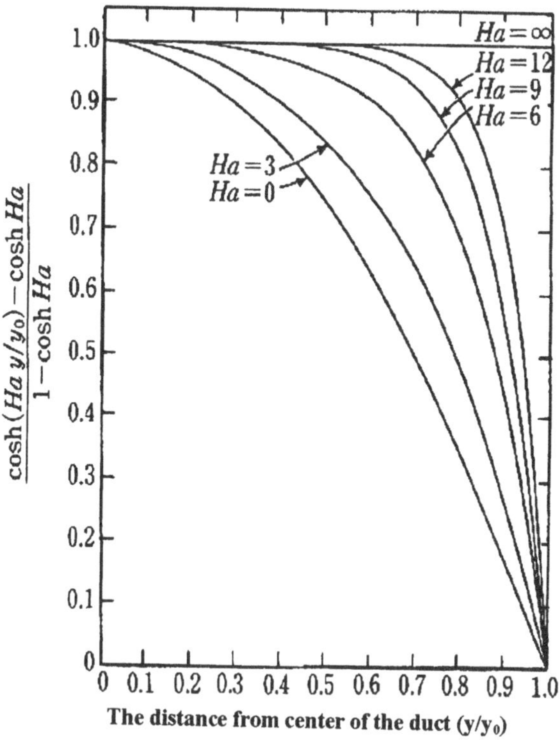

Fig. 3.7 Velocity distribution of MHD Poiseuille flow (Hartmann flow) [1]

$$v_x = \frac{\{(z_0/\sigma y_0)+R_L\}(Ha/\tanh Ha)\,(1/\mu)(\partial p/\partial x)-(Ha^2 V_g/2y_0{}^2\sqrt{\sigma\mu}\tanh Ha)}{R_L + (z_0/y_0\sigma)(Ha/\tanh Ha)} \times \frac{y_0{}^2}{Ha^2}\left\{\frac{\cosh(Ha\,y/y_0)}{\cosh Ha} - 1\right\} \tag{3.56}$$

When $V_g = 0$ and $(\partial p/\partial x) < 0$, the value of v_x becomes positive, the flow is in the $+x$-direction so that the system works like electric power. When $V_g > 0$, the generator set in the outside plays the role of a pump to accelerate the flow in the x-direction. On the other hand, when $V_g < 0$ and $(\partial p/\partial x) > 0$, the electromagnetic body force produced by V_g is against the pressure gradient.

It is known that the velocity profile between plates is given by $\{\cosh(Ha\,y/y_0) - \cosh Ha\}$ as seen from Eq. 3.56. The velocity profile is shown in Fig. 3.7 by taking $\frac{\cosh(Ha\,y/y_0)-\cosh Ha}{1-\cosh Ha}$ in the vertical axis and (y/y_0) in a horizontal axis. Here, the reason that the vertical value has been divided by $(1-\cosh Ha)$ is for normalizing, which does not influence the profile in the y-direction. A parabolic shape at $Ha = 0$ is going to change to a trapezoid with an increase in the value of Ha. It can be qualitatively understood that the larger braking force acts on the part with the faster velocity under the magnetic field so that the velocity distribution is going to become trapezoid shaped with increasing magnetic field intensity.

Problem 3.5. Derive Eq. 3.42 with the boundary conditions of $v_x = 0$ at $y = \pm y_0$.

Problem 3.6. Derive the following equation from Eq. 3.42 by approaching Ha to 0 ($B \to 0$). $v_x = (1/2\mu) \cdot (\partial p/\partial x) \cdot y_0^2 \cdot \{(y/y_0) - 1\}$

Problem 3.7. Derive Eq. 3.43

Problem 3.8. Show that at $Ha \to \infty$, the amplitude of v_x in Eq. 3.42 is reduced to $v_x \propto 1/B_0$. Furthermore, show that v_x is expressed by $v_x \propto 1/B_0^2$ when a conducting wall is replaced by an insulator wall, which means $E_z = 0$.

4 Diffusion of a Magnetic Field

4.1 Simplified Model

The diffusion equation of a magnetic field derived in the previous section is rewritten here.

$$\underbrace{\partial \boldsymbol{B}/\partial t}_{\text{unsteady state term}} = \underbrace{\nabla \times (\boldsymbol{v} \times \boldsymbol{B})}_{\text{convection term}} + \underbrace{\nu_m \nabla^2 \boldsymbol{B}}_{\text{diffusion term}} \tag{3.12}$$

The ratio of the convection term to the diffusion term is defined as the magnetic Reynolds number.

$$Re_m = \frac{\textit{convection term}}{\textit{diffusion term}} = \frac{VB/L}{\nu_m B/L^2} = LV/\nu_m = \mu_m \sigma LV$$

The name "magnetic Reynolds number" originates from the basic Reynolds number as follows:

$$Re = \frac{\textit{convection term}}{\textit{diffusion term}} = \frac{\rho VV/L}{\mu V/L^2} = LV/\nu = \rho LV/\mu$$

which is defined by the ratio of the convection term to the diffusion term in the Navier-Stokes equation given by Eq. 3.2.

$$\begin{aligned} &\underbrace{\rho\{\partial \boldsymbol{v}/\partial t}_{\text{unsteady state term}} + \underbrace{(\boldsymbol{v} \cdot \nabla)\boldsymbol{v}\}}_{\text{convection term}} = \underbrace{-\nabla p}_{\text{pressure term}} \\ &\quad + \underbrace{\mu \nabla^2 \boldsymbol{v}}_{\text{diffusion term}} + \underbrace{\boldsymbol{f}}_{\text{force term}} \end{aligned} \tag{3.2}$$

In the case of metals, $Re_m \sim 0.1\,LV$ holds because the value of $\nu_m = (1/\mu_m \sigma)$ is about $10\,m^2/s$. In many electromagnetic processes of materials,

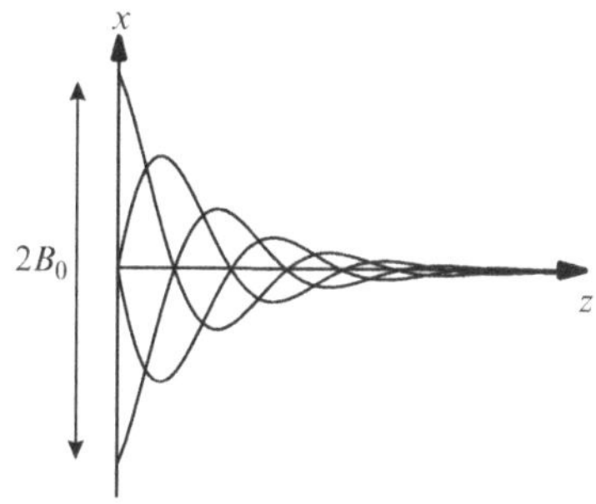

Fig. 3.8 A model of a magnetic field in a one-dimensional semi-infinite space

$Re_m < 1$ is satisfied because there are few situations in which (characteristic length $= L$) $\times$ (characteristic velocity $= V$) is over $10\,m^2/s$. Therefore, we can simplify Eq. 3.12 as follows:

$$\partial \boldsymbol{B}/\partial t = \nu_m \nabla^2 \boldsymbol{B} \tag{3.57}$$

4.2 Diffusion Model of a Magnetic Field in One-Dimensional Semi-infinite Space

The model considered here is shown in Fig. 3.8. At the surface of a metal ($z = 0$) which semi-infinitely extends in the z-direction, the magnetic field $B_0\,sin\,\omega t$ which changes with time in the x-direction is uniformly imposed on the surface (x-y plane). In this case, the component of the magnetic field is only B_x, which is the x-direction component and changes in the z-direction, so that Eq. 3.57 is reduced to

$$\partial B_x/\partial t = \nu_m \nabla^2 B_x \tag{3.58}$$

In the case where the magnetic field imposed on the surface of the metal changes periodically, B_x is also a periodic function written as follows:

$$B_x(z, t) = b_x(z)\ exp\,(i\omega t) \tag{3.59}$$

By substituting Eq. 3.59 into Eq. 3.58, the following equation is obtained.

$$i\omega b_x = \nu_m\,(d^2 b_x/dz^2) \tag{3.60}$$

Under the boundary conditions

$$b_x = B_0\ at\ z = 0, \quad b_x = 0\ at\ z = \infty \tag{3.61}$$

the solution of Eq. 3.60 is obtained as Eq. 3.62.

$$B_x\,(z,t) = B_0\; exp\left(-\sqrt{\omega/2\nu_m z}\right)\cdot exp\left\{i\left(-\sqrt{\omega/2\nu_m}\;z+\omega t\right)\right\} \tag{3.62}$$

Problem 3.9. Derive Eq. 3.62 using the relation $\sqrt{i} = (i+1)/\sqrt{2}$.

4.3 Magnetic Pressure and Joule Heat

The alternating magnetic field induces a force, a so-called magnetic pressure, and heat, so-called Joule heat, in an electrically conductive medium. Let us understand what the magnetic pressure and Joule heat are by looking at how we derive them in mathematical formulation.

By substituting Eq. 3.62 into Eq. 3.6, a current density is obtained as

$$\boldsymbol{J} = (1/\mu_m)\nabla\times\boldsymbol{B} = \frac{1}{\mu_m}\begin{vmatrix}\boldsymbol{i}_x & \boldsymbol{i}_y & \boldsymbol{i}_z\\ \dfrac{\partial}{\partial x} & \dfrac{\partial}{\partial y} & \dfrac{\partial}{\partial z}\\ B_x & 0 & 0\end{vmatrix} = \frac{1}{\mu_m}\left(0\boldsymbol{i}_x,\frac{\partial B_x}{\partial z}\boldsymbol{i}_y,-\frac{\partial B_x}{\partial y}\boldsymbol{i}_z\right)$$

Since B_x is a constant along the y-direction, the current only has the y-component, and the z-component is zero. By using B_x given in Eq. 3.62, J_y is obtained as Eq. 3.63.

$$\begin{aligned}J_y = (1/\mu_m)(\partial B_x/\partial z) &= -(1/\mu_m)\sqrt{\omega/2\nu_m}\,(1+i)\,B_0\; exp\,(-\sqrt{\omega/2\nu_m}\,z)\\ &\times exp\{i(-\sqrt{\omega/2\nu_m}\,z+\omega t)\}\end{aligned} \tag{3.63}$$

The distance in which the amplitude of B_x attenuates $1/e$ of B_0 at the surface is called skin depth δ, which is expressed as Eq. 3.64 by using Eqs. 3.62 and 3.63, and $\omega = 2\pi f$.

$$\delta = \sqrt{2\nu_m/\omega} = \sqrt{2/\mu_m\sigma\omega} = \sqrt{1/\pi f\,\mu_m\sigma} \tag{3.64}$$

Figure 3.9 shows the relationship between frequency $f(=\omega/2\pi)$ and skin depth δ of several metals.Taking cross products of $\boldsymbol{J} = (0, J_y, 0)$ and $\boldsymbol{B} = (B_x, 0, 0)$ yields an electromagnetic body force on the basis of $\boldsymbol{f} = \boldsymbol{J}\times\boldsymbol{B}$.

$$\boldsymbol{f} = \boldsymbol{J}\times\boldsymbol{B} = \begin{vmatrix}\boldsymbol{i}_x & \boldsymbol{i}_y & \boldsymbol{i}_z\\ 0 & J_y & 0\\ B_x & 0 & 0\end{vmatrix} = (0\boldsymbol{i}_x,\;0\boldsymbol{i}_y,-J_yB_x\boldsymbol{i}_z) \tag{3.65}$$

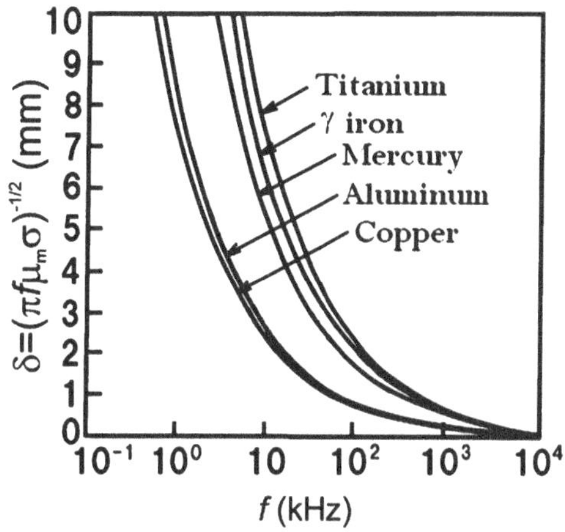

Fig. 3.9 Relationship between frequency and skin depth of typical metals

Equation 3.65 indicates that f has the z-component only.

$$f_z = -J_y B_x \tag{3.66}$$

Generally speaking, when an imposed magnetic field frequency is over 100 Hz, the motion of a liquid metal cannot follow with such a high frequency of the electromagnetic body force. Thus, the time-averaged value of the electromagnetic body force reduces to a substantial driving force acting on liquid metals. The time averaged force per period is expressed as follows:

$$\bar{f_z} = (1/T)\int_0^T f_z dt = (1/T)\int_0^T (-J_y B_x)dt = (1/2)\,Re\{-j_y \cdot b_x^*\}, \tag{3.67}$$

where T is the period of a periodic function and Re{} indicates the real part of {}. Here j_y and b_x are defined by the functions of $J_y = j_y\ exp\ i\omega t, B_x = b_x\ exp\, i\omega t$ and b_x^* expresses the complex conjugate number of b_x.

Let us show a concise procedure to derive the mean time value of the product in two variables with periodic change. Namely, the time-averaged value of the product of two variables can be expressed as follows:

$$\begin{aligned}
&(1/T)\int_0^T Re(A\cdot exp\ i\omega t)\cdot Re\,(B\cdot exp\ i\omega t)dt \\
&= (1/T)\int_0^T Re\,\{(a_1 + a_2 i)\cdot(\cos\,\omega t + i\ sin\ \omega t)\} \\
&\quad \times Re\,\{(b_1 + b_2 i)\cdot(\cos\ \omega t + i\,sin\ \omega t)\}dt
\end{aligned}$$

$$= (1/T)\int_0^T (a_1\cos\omega t - a_2\, sin\,\omega t)\times(b_1\cos\omega t - b_2\, sin\,\omega t)dt$$

$$= (1/T)\int_0^T (a_1\; b_1\cos^2\omega t dt + (1/T)\int_0^T a_2 b_2\, sin^2\,\omega t dt$$

$$\underbrace{-(1/T)\int_0^T (a_1 b_2 + a_2 b_1) sin\,\omega t\cdot\cos\,\omega t)dt}_{\substack{\| \\ 0}}$$

$$= \frac{1}{2}(a_1 b_1 + a_2 b_2) \tag{3.68}$$

In the derivation, the following relations have been used.

$$cos^2\,\omega t = (1+\cos 2\omega t)/2 \text{ and } sin^2\,\omega t = (1-\cos 2\omega t)/2$$

Here $A = a_1 + a_2 i$, $B = b_1 + b_2 i$, (a_1, a_2, b_1 and b_2 are real numbers.) Thus, $Re\,(A^*\cdot B) = Re\{(a_1 - a_2 i)(b_1 + b_2 i)\} = a_1 b_1 + a_2 b_2$, and $Re(A\cdot B^*) = Re\{(a_1 + a_2 i)(b_1 - b_2 i)\} = a_1 b_1 + a_2 b_2$,

$$(1/T)\int_0^T Re\,(A\cdot\exp\, i\omega t)\cdot Re(B\cdot exp\, i\omega t)\,dt$$
$$= Re(A^*\cdot B)/2 = Re(A\cdot B^*)/2 \tag{3.69}$$

Then, Eq. 3.67 has been proved.

$\bar{f_z}$ is obtained by substituting Eqs. 3.62 and 3.63 into Eq. 3.67.

$$\bar{f_z} = (1/2)Re(-j_y b_x^*)$$
$$= (1/2)Re\left[-\left(\frac{-1}{\mu_m}\right)\sqrt{\omega/2\nu_m}(1+i)B_0\, exp(-\sqrt{\omega/2\nu_m}z)\exp\, i\left(-\sqrt{\omega/2\nu_m}z\right)\right.$$
$$\left.\times B_0\, exp(-\sqrt{\omega/2\nu_m}z)exp\left\{-i(-\sqrt{\omega/2\nu_m}z)\right\}\right]$$
$$= ({B_0}^2/2\delta\mu_m)\, exp\,(-2z/\delta) \tag{3.70}$$

The positive value of Eq. 3.70 means that a body force acting from the surface to the inside (positive direction of the z-axis) arises by imposing a magnetic field on the surface of a metal. The force distribution also decreases from the surface toward the inside as B_x and J_x do. The degree of the force decay is two times faster than those of B_x and J_x, namely the force decays to the value of 1/e at 1/2 distance of the skin depth δ, i.e. at $z = \delta/2$. That is, when an alternating magnetic field is imposed on a metal, the force is very concentrated in the vicinity of the metal surface.

Integrating $\bar{f_z}$ from the surface ($z = 0$) to an infinite distance ($z = \infty$) yields Eq. 3.71.

$$p_m = \int_0^\infty \bar{f}_z dz = \int_0^\infty (B_0{}^2/2\delta\mu_m) exp(-2z/\delta) dz$$
$$= B_0^2/4\mu_m = B_e^2/2\mu_m, \tag{3.71}$$

where, B_e $(= B_0/\sqrt{2})$ means the effective value of the magnetic field.

The reason why the integrated value of $\bar{f}_z$ is considered to be a pressure on the surface is explained as followings. The MKS unit of an electromagnetic body force $\bar{f}_z$ is (N/m^3). Because the force concentrates just in the vicinity of the molten metal surface, p_m obtained from Eq. 3.71 can be regarded as the pressure acting on the metal surface. Here, the unit of p_m is $(N/m^2) = Pa$ and the value of p_m is positive regardless of plus and minus signs of B_e as seen from Eq. 3.71. That is, p_m always acts to push the metal surface down and is called a magnetic pressure. The time averaged heat generation rate per unit volume can be obtained by using Eqs. 3.63 and 3.69.

$$\bar{q} = (1/T) \int_0^T J_y^2/\sigma \, dt = Re(j_y^* \, j_y)/2\sigma = \left(\omega B_0^2/2\mu_m\right) exp \left(-2\sqrt{\omega/2\nu_m}\, z\right) \tag{3.72}$$

Since the heat generated by $\bar{q}$ diffuses and disappears, the $\bar{q}$ is called an Ohmic dissipation. In addition, the Ohmic dissipation per unit area is given by Eq. 3.73.

$$Q = \int_0^\infty \bar{q} dz = (1/2\sqrt{2})\,(1/\mu_m)\sqrt{(\omega/\mu_m\sigma)}\,B_0^2 \tag{3.73}$$

The relationship between p_m and Q is obtained by using Eqs. 3.71 and (3.73).

$$Q = p_m\sqrt{2\omega/\mu_m\sigma} \tag{3.74}$$

Problem 3.10. Derive Eq. 3.71.

Problem 3.11. Derive Eq. 3.72.

Problem 3.12. Derive Eq. 3.74.

4.4 Diffusion Model of a Magnetic Field in Limited One-Dimension

This problem is simplified by neglecting fringing effects, which appear on both edges of a conductor. Let us consider a conductor with infinite length set in a solenoid coil with infinite length. As shown in Fig. 3.10, an alternating magnetic field is generated by passing an alternating current in a coil. Under this condition

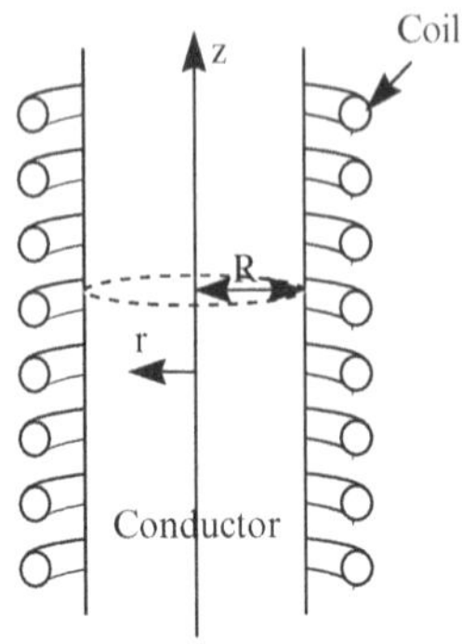

Fig. 3.10 A model of a magnetic field penetrating into a cylindrical conductor

a distribution of the magnetic field penetrating into the conductor exists. Since the coil and conductor are infinite, the magnetic field only has the component of axial direction (*z*-direction).

Assuming an axial symmetry, the diffusion equation of the magnetic field is described as the following equation.

$$(1/r)\,\partial\{r(\partial B_z/\partial r)\}/\partial r = \sigma \mu_m\, \partial B_z/\partial t \tag{3.75}$$

Here, B_z is a periodic function and expressed by Eq. 3.76.

$$B_z\,(r,t) = b_z\,(r)\, exp\,(i\omega t) \tag{3.76}$$

Substituting Eq. 3.76 into Eq. 3.75 and rearranging the resulting equation yields Eq. 3.77.

$$d^2b_z/dr^2 + (1/r)\, db_z/dr - i\omega\sigma\, \mu_m\, b_z = 0 \tag{3.77}$$

Here, by designating $k^2 = -i\omega\sigma\, \mu_m,\ x = k\, r$,

$$d^2b_z/dx^2 + (1/x)db_z/dx + b_z = 0 \tag{3.78}$$

Equation 3.78 is a Bessel's equation of zero order, and its solution is given by Eq. 3.79.

$$b_z = C_1 J_0(x) + C_2 Y_0(x) \tag{3.79}$$

Here, $C_2 = 0$ because of $Y_0\,(0) = -\infty$. Thus, Eq. 3.79 becomes as follows:

$$b_z = C_1\, J_0(x) \tag{3.80}$$

Now, putting $b_z = B_0$ at $r = R$ (conductor surface), the solution of the magnetic field distribution in the conductor is given as follows:

$$B_z = B_0\{J_0(kr)/J_0(kR)\}\, exp\,(i\omega t) \tag{3.81}$$

However, we cannot immediately get the magnetic field distribution (real part) from Eq. 3.81, because k appearing in Eq. 3.81 is a complex number as seen in the definition of $k^2 = -i\omega\sigma\mu_m$. Thus, let us separate the real and imaginary parts of Eq. 3.81 by introducing the Kelvin function. The relation between the Kelvin function of order ν,$\{ber_\nu(z), bei_\nu(z)\}$ and the Bessel function $J_\nu(z)$ is written as follows:

$$ber_\nu\,(z) \pm i\,bei_\nu(z) = J_\nu\,(e^{\pm 3\pi i/4} z) \tag{3.82}$$

The coefficient k introduced in Eq. 3.78 can be expressed by Eq. 3.83.

$$k = \sqrt{-i\omega\sigma\mu_m} = e^{\pm 3\pi i/4}\,\sqrt{\omega\sigma\nu_m} \tag{3.83}$$

By using Eqs. 3.82 and 3.83, the magnetic field distribution B_z in Eq. 3.81 is expressed as follows:

$$\begin{aligned} B_z &= B_0\,[\{ber_0(nr) + i\ bei_0(nr)\}/\{ber_0(nR) + i\ bei_0\,(nR)\}]\,e^{i\omega t} \\ &= B_0\frac{\{ber_0(nr) + i\ bei_0(nr)\}\{ber_0(nR) - i\ bei_0(nR)\}}{\{ber_0\,(nR)\}^2 + \{bei_0(nR)\}^2}e^{i\omega t} \end{aligned} \tag{3.84}$$

Here, $n = \sqrt{\omega\sigma\mu_m}$.

The magnetic field distribution, which is obtained from the real part of Eq. 3.84 is given as Eq. 3.85.

$$\begin{aligned} B_z = B_0[&\{ber_0(nr)ber_0(nR) + bei_0(nr)bei_0(nR)\}\cos\,\omega t \\ &- \{bei_0(nr)ber_0(nR) - ber_0(nr)bei_0(nR)\}sin\,\omega t]/ \\ &\left[\{ber_0(nR)\}^2 + \{bei_0(nR)\}^2\right] \end{aligned} \tag{3.85}$$

On the other hand, an electric current density J_θ is derived from Ampere's law (Eq. 3.6) as Eq. 3.86.

$$J_\theta = (1/\mu_m)(-\partial B_z/\partial r) \tag{3.86}$$

J_θ is obtained by substituting Eq. 3.81 into Eq. 3.86 and using the relation of $\sqrt{-i} = (i-1)/\sqrt{2}$.

$$\begin{aligned} J_\theta &= (-1/\mu_m)\{B_0/J_o(kR)\}\{dJ_0(kr)/dr\}e^{i\omega t} = (B_0k/\mu_m)\{J_1(kr)/J_0(kR)\}e^{i\omega t} \\ &= \{B_0(i-1)n/\sqrt{2}\mu_m\}[\{ber_1(nr) + i\ bei_1\,(nr)\}/\{ber_0(nR) + i\ bei_0(nR)\}]e^{i\omega t} \end{aligned} \tag{3.87}$$

The real part of Eq. 3.87 is expressed as Eq. 3.88.

$$\begin{aligned} J_\theta = {} & B_0 n / (\sqrt{2}\mu_m[\{ber_0(nR)\}^2 + \{bei_0(nR)\}^2]) \\ & \times [-\{ber_1(nr)ber_0(nR) + bei_1(nr)bei_0(nR)\}\{\cos \omega t + sin\, \omega t\} \\ & + \{bei_1(nr)ber_0(nR) - ber_1(nr)bei_0(nR)\}\{sin\, \omega t - \cos \omega t\}] \end{aligned} \tag{3.88}$$

A Lorentz force $\boldsymbol{f}$ acting on an electric conductor is obtained by substituting $B = (0, 0, B_z)$ and $J = (0, J_\theta, 0)$ into Eq. 3.17.

$$\boldsymbol{f} = \begin{vmatrix} \boldsymbol{i}_r & \boldsymbol{i}_\theta & \boldsymbol{i}_z \\ 0 & J_\theta & 0 \\ 0 & 0 & B_z \end{vmatrix} = (J_\theta B_z \boldsymbol{i}_r,\ 0\boldsymbol{i}_\theta,\ 0\boldsymbol{i}_z) \tag{3.89}$$

From Eq. 3.89, it can be understood that the force only works in the radius direction. The time average value of the force over a period $\bar{f}_r$ is derived from Eq. 3.69.

$$\bar{f}_r = Re(j_\theta \cdot b_z^*)/2 \tag{3.90}$$

Here, $j_\theta(r)$ is defined as $J_\theta(r,t) \equiv j_\theta(r)\, e^{i\omega t}$ and b_z^* is the complex conjugate number of b_z. Substituting Eqs. 3.84 and 3.87 into Eq. 3.90 yields Eq. 3.91 for the time average value of the force acting on the conducting materials.

$$\begin{aligned} \bar{f}_r &= \frac{1}{2} Re\left[\frac{B_0(i-1)n}{\sqrt{2}\mu_m} \cdot \frac{\{ber_1(nr) + i\ bei_1(nr)\}}{\{ber_0(nR) + i\ bei_0(nR)\}} \times \frac{B_0\{ber_0(nr) - i\ bei_0(nr)\}}{\{ber_0(nR) - i\ bei_0(nR)\}} \right] \\ &= -\frac{B_0^2\sqrt{\omega\sigma}}{2\sqrt{2}\mu_m} \cdot \frac{ber_0(nr)\{ber_1(nr) + bei_1(nr)\} + bei_0(nr)\{bei_1(nr) - ber_1(nr)\}}{\{ber_0(nR)\}^2 + \{bei_0(nR)\}^2} \end{aligned} \tag{3.91}$$

4.5 Traveling Magnetic Field [2, 3]

As shown in Fig. 3.11, in order to transport a metal in the x-direction, an alternating magnetic field propagating in the x-direction, which is described by Eq. 3.92 is imposed by use of a linear motor.

$$\boldsymbol{B}(x,z,t) = \boldsymbol{b}(x,z) \cdot exp\,(i\omega t) \tag{3.92}$$

Equation 3.93 is obtained by substituting Eq. 3.92 into the diffusion equation of the magnetic field expressed by Eq. 3.57.

$$i\omega \boldsymbol{b} = \nu_m \nabla^2 \boldsymbol{b} \tag{3.93}$$

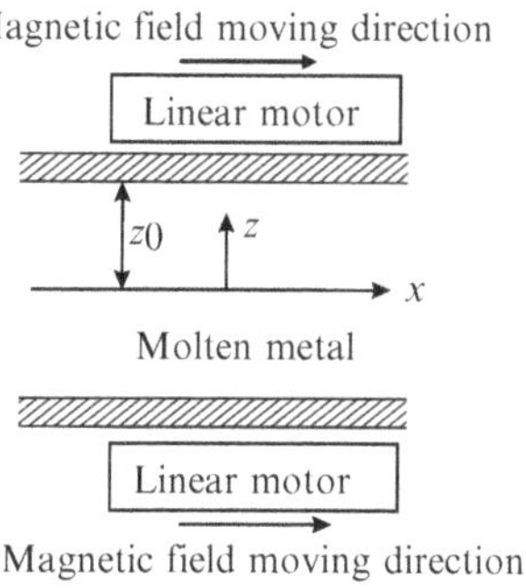

Fig. 3.11 A frame format of equipment where a metal is driven by a traveling magnetic field

The magnetic field propagates as a sinusoidal wave in the x-direction with a propagation constant γ and attenuates only in the direction of the metal depth (z-direction). This magnetic field is described as the following function:

$$\boldsymbol{b}(x,z) = \begin{cases} b_x = \hat{b}_x(z) \cdot exp\,(-i\gamma x) \\ b_y = 0 \\ b_z = \hat{b}_z(z) \cdot exp\,(-i\gamma x) \end{cases} \tag{3.94}$$

Substituting b_x which is the x-component of Eq. 3.94 into Eq. 3.93 and rearranging the resulting equation yields Eq. 3.95.

$$d^2\hat{b}_x/dz^2 - \beta^2\hat{b}_x = 0, \tag{3.95}$$

where $\beta^2 \equiv \gamma^2 + i\omega/\nu_m$

Under the boundary condition of

$$\hat{b}_x = b_0 \text{ at } z = z_0 \text{ and } d\hat{b}_x/dz = 0 \text{ at } z = 0 \tag{3.96}$$

Equation 3.97 is obtained by solving Eq. 3.95.

$$B_x = b_0\{\cosh(\beta z)/\cosh(\beta z_0)\}\, exp\{i(\omega t - \gamma x)\} \tag{3.97}$$

The part in Eq. 3.97, $exp\{i(\omega t - \gamma x)\}$, represents a sinusoidal wave, which moves in the x-direction with a propagation constant γ.

From Eq. 3.7 which expresses Gauss' law representing the continuity of a magnetic field, Eq. 3.98 can be derived.

$$(\partial B_x/\partial x) + (\partial B_z/\partial z) = 0 \tag{3.98}$$

Solving the equation resulting from substituting Eq. 3.97 into Eq. 3.98 under the boundary condition of Eq. 3.99 yields Eq. 3.100.

$$B_z = 0 \quad \text{(symmetry) at } z = 0 \tag{3.99}$$

$$B_z = i\gamma b_0 \{\sinh(\beta z)/\beta \cosh(\beta z_0)\} \exp\{i(\omega t - \gamma x)\} \tag{3.100}$$

Next, the current density is obtained by substituting the relation given by Eq. 3.10, Eqs. 3.97 and 3.100 into Eq. 3.6.

$$J = \begin{cases} J_x = 0 \\ J_y = (b_0/\mu_m)(\beta^2 - \gamma^2)\{\sinh(\beta z)/\beta \cosh(\beta z_0)\} \exp\{i(\omega t - \gamma x)\} \\ J_z = 0 \end{cases} \tag{3.101}$$

Finally, by substituting Eqs. 3.97, 3.100 and 3.101 into Eq. 3.17 and using the relation of Eq. 3.69, the time average value of Lorentz force $\boldsymbol{f}$ is reduced as follows:

$$\bar{\boldsymbol{f}} = \begin{cases} \bar{f}_x = \dfrac{b_0^2 \omega \sigma \gamma}{2\sqrt{\gamma^4 + (\mu_m \omega \sigma)^2}} \cdot \dfrac{\cosh(2\beta_r z) - \cos(2\beta_i z)}{\cosh(2\beta_r z_0) + \cos(2\beta_i z_0)} \\ \bar{f}_y = 0 \\ \bar{f}_z = \dfrac{b_0^2 \omega \sigma}{2\sqrt{\gamma^4 + (\mu_m \omega \sigma)^2}} \cdot \dfrac{-\beta_i \sinh(2\beta_r z) + \beta_r sin(2\beta_i z)}{\cosh(2\beta_r z_0) + \cos(2\beta_i z_0)} \end{cases} \tag{3.102}$$

Here, β_r and β_i represent the real and imaginary parts of β, respectively.

The distribution of the magnetic field vector at the instance of t = 0 is shown in Fig. 3.12, which was calculated by use of Eqs. 3.97 and 3.100 under the conditions of $\gamma = 1, \omega \ll \gamma^2 v_m, z_0 = 1$. This magnetic field has a z-axis symmetry and its z-component is zero on the z-axis.

Another type of traveling magnetic field is obtained as Eqs. 3.103 and 3.104 by solving Eqs. 3.95 and 3.98 under the conditions given in Eq. 3.105.

$$B_x = b_0 \{\sinh(\beta z)/\sinh(\beta z_0)\} exp\{i(\omega t - \gamma x)\} \tag{3.103}$$

$$B_z = i\gamma b_0 \{\cosh(\beta z)/\beta \sinh(\beta z_0)\} exp\{i(\omega t - \gamma x)\} \tag{3.104}$$

$$\hat{b}_x = b_0 \text{ at } z = z_0, \text{ and } d\hat{b}_z/dz = 0 \text{ and } \hat{b}_x = 0 \text{ at } z = 0 \tag{3.105}$$

In this case the distribution of the magnetic field vector at the instance of t = 0 is shown in Fig. 3.13, which was calculated under the condition of $\gamma = 1, \omega \ll \gamma^2 v_m, z_0 = 1$ We can see that the magnetic field crosscuts the z-axis. This point is an important difference between the magnetic fields shown in Figs. 3.12 and 3.13.

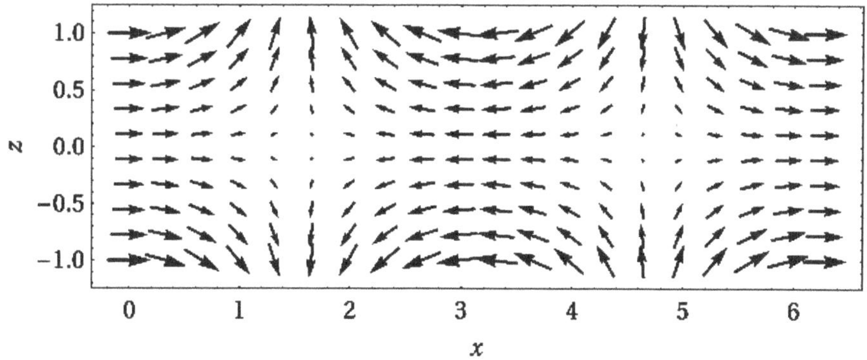

Fig. 3.12 The distribution of a traveling magnetic field with z-axis symmetry at t = 0, calculated under the condition of $\gamma = 1$, $\omega \ll \gamma^2 \nu_m$ and $z_0 = 1$

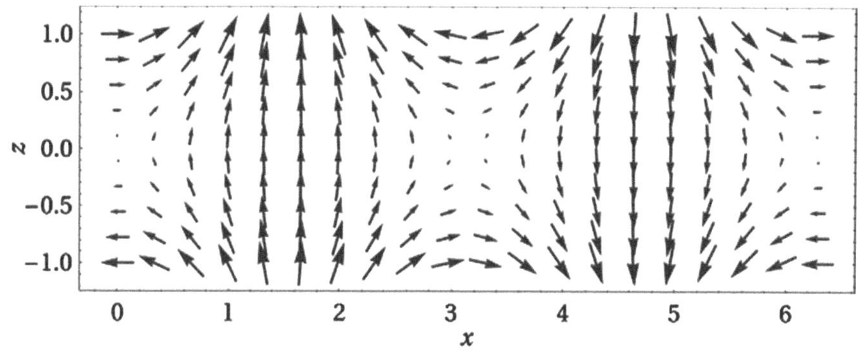

Fig. 3.13 The distribution of a traveling transverse magnetic field at t = 0, calculated under the condition of $\gamma = 1$, $\omega \ll \gamma^2 \nu_m$ and $z_0 = 1$

The current density is obtained as Eq. 3.106 from Eqs. 3.103 and 3.104.

$$\boldsymbol{J} = \begin{cases} J_x = 0 \\ J_y = (b_0/\mu_m)(\beta^2 - \gamma^2)\{\text{conh}(\beta z)/\beta \sinh(\beta z_0)\} \exp\{i(\omega t - \gamma x)\} \\ J_z = 0 \end{cases} \tag{3.106}$$

Then, the time average value of the Lorentz force is reduced as follows:

$$\bar{f} = \begin{cases} \bar{f}_x = \dfrac{b_0^2 \omega \sigma \gamma}{2\sqrt{\gamma^4 + (\mu_m \omega \sigma)^2}} \cdot \dfrac{\cosh(2\beta_r z) + \cos(2\beta_i z)}{\cosh(2\beta_r z_0) - \cos(2\beta_i z_0)} \\ \bar{f}_y = 0 \\ \bar{f}_z = -\dfrac{b_0^2 \omega \sigma}{2\sqrt{\gamma^4 + (\mu_m \omega \sigma)^2}} \cdot \dfrac{\beta_i \sinh(2\beta_r z) + \beta_r \sin(2\beta_i z)}{\cosh(2\beta_r z_0) - \cos(2\beta_i z_0)} \end{cases} . \quad (3.107)$$

Let us compare the two types of traveling magnetic field given in Figs. 3.12 and 3.13 by taking the case of $|\beta| z_0 \sim 1$ from a practical view point. The driving force in the x-direction $\bar{f}_x$ approaches zero value toward $z = 0$ in the magnetic field given in Fig. 3.12, which is easily understood from Eq. 3.102. On the other hand, the $\bar{f}_x$ given in Eq. 3.107 does not take a zero value at $z = 0$. Thus, the transverse type magnetic field shown in Fig. 3.13 is recommended in the practical use of linear motors and electromagnetic pumps.

5 Boundary Conditions

The conditions of velocity, stress and electromagnetic field to be satisfied at the boundary between fluid (I) and fluid (II) are described as follows:

(a) **Velocity**

$$\boldsymbol{v}^{(I)} - \boldsymbol{v}^{(II)} = 0 \quad (3.108)$$

Equation 3.108 indicates $v_x^{(I)} = v_x^{(II)}, v_y^{(I)} = v_y^{(II)}$ and $v_z^{(I)} = v_z^{(II)}$. As shown in Fig. 3.14, the superscripts of (I) and (II) represent the fluid (I) and fluid (II), respectively. When the shape of the boundary surface between fluid (I) and fluid (II) is given by the equation

$$F(\boldsymbol{r}, t) = 0, \quad (3.109)$$

the moving velocity of the boundary surface DF/Dt observed from the coordinates traveling with velocity $\boldsymbol{v}$ must become zero. That is, the variation of the shape of the boundary surface can not be observed from the coordinates traveling together with the boundary surface. This matter is expressed mathematically by $DF/Dt = 0$. Therefore, the following equation is obtained.

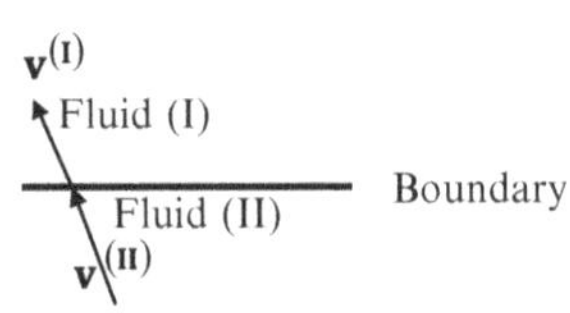

Fig. 3.14 Velocity continuity at the boundary between two fluids

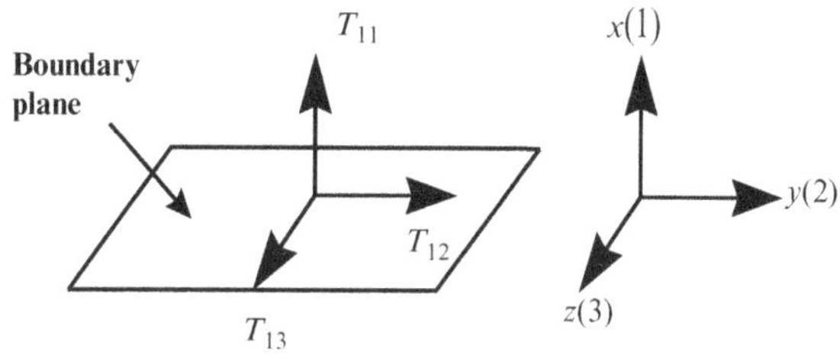

Fig. 3.15 Stress components at the boundary plane

$$DF/Dt = \partial F/\partial t + (\boldsymbol{v} \cdot \nabla)F = 0 \tag{3.110}$$

(b) **Stress**

The boundary condition of stress is given by Eq. 3.111.

$$\left|\boldsymbol{n} \cdot (\boldsymbol{T}^{(I)} - \boldsymbol{T}^{(II)})\right| = 0, \tag{3.111}$$

where $\boldsymbol{n}$ is a unit normal vector perpendicular to the boundary plane. A scalar (dot) product of the vector $\boldsymbol{n}$ and the tensor $\boldsymbol{T}$ is described as follows:

$$|\boldsymbol{n} \cdot \boldsymbol{T}| = (1, 0, 0) \begin{pmatrix} T_{11} & T_{12} & T_{13} \\ T_{21} & T_{22} & T_{23} \\ T_{31} & T_{32} & T_{33} \end{pmatrix} = (T_{11}, T_{12}, T_{13})$$

For instance, the tensors of T_{11}, T_{12} and T_{13} can be imagined as shown in Fig. 3.15. Then, Eq. 3.111 is described as $T_{11}^{(I)} = T_{11}^{(II)}, T_{12}^{(I)} = T_{12}^{(II)}$, and $T_{13}^{(I)} = T_{13}^{(II)}$. In addition, when a surface tension acts on the boundary surface, Eq. 3.112 is obtained as

$$\boldsymbol{n} \cdot \left|\boldsymbol{n} \cdot (\boldsymbol{T}^{(I)} - \boldsymbol{T}^{(II)})\right| = 2\sigma_f / r \tag{3.112}$$

Here, σ_f and r are a surface tension and an average curvature radius, respectively.

As $\boldsymbol{n} = (1, 0, 0)$ and $\left|\boldsymbol{n} \cdot (\boldsymbol{T}^{(I)} - \boldsymbol{T}^{(II)})\right| = (\Delta T_{11}, \Delta T_{12}, \Delta T_{13})$ hold, Eq. 3.112 can be expressed as

$$\Delta T_{11} = T_{11}^{(I)} - T_{11}^{(II)} = 2\sigma_f / r$$

While, a tangential component on the boundary surface is given as Eq. 3.113.

$$\boldsymbol{n} \times \left|\boldsymbol{n} \cdot (\boldsymbol{T}^{(I)} - \boldsymbol{T}^{(II)})\right| = 0 \tag{3.113}$$

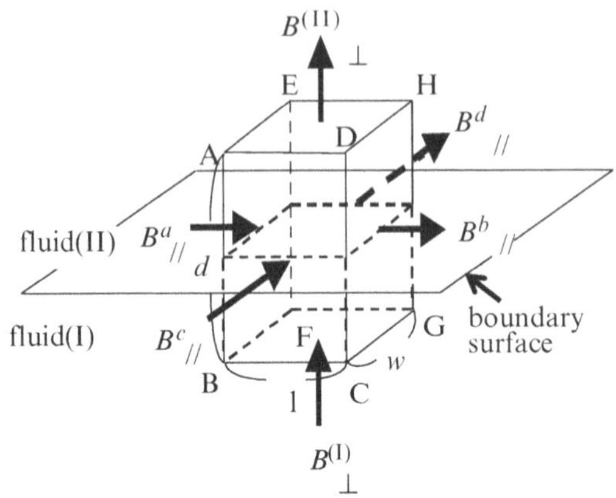

Fig. 3.16 Magnetic flux components applied to a volume ABCDEFGH containing a boundary surface

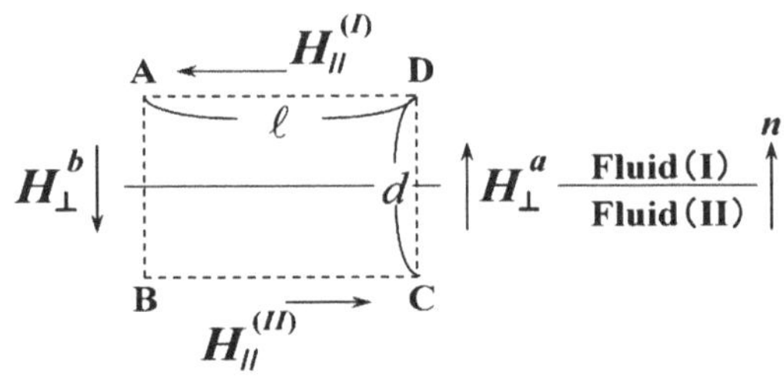

Fig. 3.17 Magnetic fields applying a surface ABCD containing a boundary surface

That is, Eq. 3.113 can be described as

$$\Delta T_{12} = T_{12}^{(I)} - T_{12}^{(II)} = 0$$
$$\Delta T_{13} = T_{13}^{(I)} - T_{13}^{(II)} = 0,$$

(c) **Electric and magnetic fields**

Conservation of a magnetic field Let us image a volume ABCDEFGH containing a boundary surface as shown in Fig. 3.16. Applying $\nabla \cdot \boldsymbol{B} = 0$ (the conservation of a magnetic field) to the volume yields the following equation.

$$\int \nabla \cdot \boldsymbol{B} dv = lw\left(B_{\perp}^{(I)} - B_{\perp}^{(II)}\right) + wd\left(B_{\parallel}^{a} - B_{\parallel}^{b}\right) + ld\left(B_{\parallel}^{c} - B_{\parallel}^{d}\right) = 0$$

By reducing d to 0, $B_{\perp}^{(I)} = B_{\perp}^{(II)}$ is obtained. Namely, Eq. 3.114 is obtained.

$$\boldsymbol{n} \cdot (\boldsymbol{B}^{(I)} - \boldsymbol{B}^{(II)}) = 0 \tag{3.114}$$

Ampere's law Let us image a surface ABCD containing a boundary surface as shown in Fig. 3.17 and apply $\nabla \times \boldsymbol{H} = \boldsymbol{J}$ to it, then the following equation is obtained.

$$\int \nabla \times \boldsymbol{H} d\boldsymbol{S} = \oint \boldsymbol{H} d l = l\left(H_{\|}^{(II)} - H_{\|}^{(I)}\right) + d\left(H_{\perp}^{a} - H_{\perp}^{b}\right) = \int \boldsymbol{J} d\boldsymbol{S} = J_s l$$

By reducing d to 0, $H_{\|}^{(II)} - H_{\|}^{(I)} = J_s$ is obtained. Here, J_s is a current per unit length on the boundary surface that flows in the direction perpendicular to the plane of a cross-section with infinitely small thickness. J_s is called a surface current.

$$\boldsymbol{n} \times (\boldsymbol{H}^{(I)} - \boldsymbol{H}^{(II)}) = \boldsymbol{J}_s \tag{3.115}$$

Conservation of charge By applying $\nabla \cdot \boldsymbol{D} = \rho_e$ (conservation law of electric charge) to an interface as well as the derivation of Eq. 3.114, Eq. 3.116 can be derived.

$$\boldsymbol{n} \cdot (\boldsymbol{D}^{(I)} - \boldsymbol{D}^{(II)}) = \rho_{es} \tag{3.116}$$

Here, ρ_{es} is an electric charge appearing on a unit interface.

Electromagnetic induction Let us apply Faraday's law ($\nabla \times \boldsymbol{E} = -\partial \boldsymbol{B}/\partial t$) to the interface as well as the derivation of Eq. 3.115. In this case, the magnetic flux passing through the plane of ABCD approaches zero when d is getting smaller, i.e., $\partial \boldsymbol{B}/\partial t = 0$. Thus, Eq. 3.117 is derived.

$$\boldsymbol{n} \times (\boldsymbol{E}^{(II)} - \boldsymbol{E}^{(II)}) = 0 \tag{3.117}$$

Let us list all of the equations mentioned above.

From $\boldsymbol{n} \cdot (\boldsymbol{B}^{(I)} - \boldsymbol{B}^{(II)}) = 0$ (3.114), $B_{\perp}^{(I)} = B_{\perp}^{(II)}$ (3.118)

From $\boldsymbol{n} \times (\boldsymbol{H}^{(I)} - \boldsymbol{H}^{(II)}) = \boldsymbol{J}_s$ (3.115), $H_{\|}^{(I)} - H_{\|}^{(II)} = J_s$ (3.119)

From $\boldsymbol{n} \cdot (\boldsymbol{D}^{(I)} - \boldsymbol{D}^{(II)}) = \rho_{\mathrm{es}}$ (3.116), $D_{\perp}^{(I)} - D_{\perp}^{(II)} = \rho_{\mathrm{es}}$ (3.120)

From $\boldsymbol{n} \times (\boldsymbol{E}^{(I)} - \boldsymbol{E}^{(II)}) = 0$ (3.117), $E_{\|}^{(I)} = E_{\|}^{(II)}$ (3.121)

6 Dimensionless Numbers Relating to Magnetohydrodynamics

The governing equations that describe phenomena relating to fluid motion and electromagnetic fields are the momentum balance equation of Eq. 3.2 and the diffusion equation of magnetic field of Eq. 3.12.

$$\underbrace{\rho\,\{\partial \boldsymbol{v}/\partial t}_{\substack{\text{unsteady}\\ \text{state}\\ \text{term}\\ (A-I)}} + \underbrace{(\boldsymbol{v}\cdot\nabla)\boldsymbol{v}\}}_{\substack{\text{convection}\\ \text{term}\\ (A-II)}} = \underbrace{\mu\nabla^2\boldsymbol{v}}_{\substack{\text{diffusion}\\ \text{term}\\ (A-III)}} + \underbrace{\boldsymbol{J}\times\boldsymbol{B}}_{\substack{\text{force term}\\ (A-IV)}} \quad \underbrace{\nabla p}_{\substack{\text{pressure}\\ \text{term}\\ (A-V)}} \tag{3.2}$$

$$\underbrace{\partial \mathbf{B}/\partial t}_{\substack{\text{unsteady}\\ \text{state}\\ \text{term}\\ (B-I)}} - \underbrace{\nabla\times(\boldsymbol{v}\times\boldsymbol{B})}_{\substack{\text{convection}\\ \text{term}\\ (B-II)}} = \underbrace{\nu_m\nabla^2\boldsymbol{B}}_{\substack{\text{diffusion}\\ \text{term}\\ (B-III)}} \tag{3.12}$$

Taking the ratio of each term, dimensionless numbers can be derived.

$$Ha = \sqrt{(A-IV)/(A-III)} = \sqrt{(JB)/(\mu V/L^2)} = \sqrt{\{(\sigma VB)\cdot B\}/(\mu V/L^2)}$$
$$= BL\sqrt{\sigma/\mu} \qquad (\textit{Hartmann number})$$

$$N = (A-IV)/(A-II) = (JB)/(\rho V^2/L) = \{(\sigma VB)\cdot B\}/(\rho V^2/L) = \sigma \mathrm{B}^2 L/\rho V$$

(*Stuart number*) or (*Interaction parameter*)

$$Re_m = (B-II)/(B-III) = (VB/L)/(\nu_m B/L^2) = VL/\nu_m$$

(*magnetic Reynolds number*)

$$W_m = \sqrt{(B-I)/(B-III)} = \sqrt{(\omega B)/(\nu_m B/L^2)} = L\sqrt{\omega/\nu_m} = L'/\delta$$

(*Shielding parameter*) or (*magnetic Womersley number*)

where L' is another characteristic length.

Other dimensionless numbers appearing in magnetohydrodynamic phenomena are,

$$Ma_m = (\textit{fluid velocity})/(\textit{Alfvén velocity})$$
$$= V/\sqrt{B^2/\mu_m\rho} \qquad (\textit{magnetic Mach number})\ or\ (\textit{Alfvén number})$$

$$Pr_m = (\textit{kinematic viscosity})/(\textit{diffusion coefficent of magnetic field})$$
$$= \{(\mu/\rho)/(1/\sigma\mu_m)\} = \nu/\nu_m \qquad (\textit{magnetic Prandtl number})$$

$$P_m = (\text{magnetic pressure})/(\text{dynamic pressure}) = \{(B^2/\mu_m)/\rho v^2\} = B^2/\rho\mu_m V^2$$

(*magnetic pressure number*)

The physical meaning of each dimensionless number will be described in the following.

Hartmann number $Ha = BL\sqrt{\sigma/\mu} = \sqrt{ReRem/Ma_m^2}$ represents the ratio of the Lorentz force to the viscous force. In this derivation, the relation of $J \approx \sigma VB$ has been used for evaluating the induced current.

Stuart number **(or** ***Interaction parameter*****)** $N = \sigma B^2 L/\rho V$ represents the ratio of the Lorentz force term to the convection term. In this derivation, the Lorentz force term is evaluated by using $J \approx \sigma VB$, under the assumption of $Re_m \ll 1$.

magnetic Reynolds number $Re_m = VL/\nu_m = VL\sigma\mu_m$ represents the ratio of the convection term to the diffusion term of a magnetic field. An induced magnetic field B_i is reduced to $B_i \approx J\mu_m L$ from $\nabla \times \boldsymbol{H} = \boldsymbol{J}$, and $J \approx \sigma V\ B$ is obtained by Ohm's law. Here, B is the imposed magnetic field. Thus, $B_i/B \approx VL\sigma\mu_m$ is obtained. Thus, Re_m means the ratio of the induction magnetic field to the imposed magnetic field. As mentioned before, the condition of $Re_m \leq 1$ is satisfied in most processes seen in the Electromagnetic Processing of Materials.

Shielding parameter $W_m = L\sqrt{\omega/\nu_m} = L'/\delta$ is a dimensionless number representing the ratio of the changing velocity of a magnetic field to a magnetic diffusing velocity. The condition of $W_m \ll 1$ indicates that the magnetic field penetrates deeply inside a fluid, and the skin effect does not appear.

Alfvén number and magnetic Mach number In highly conductive fluids, there exists a transverse wave called an *Alfvén* wave, which is accompanied by a motion of the fluid and a fluctuation of the magnetic field. An *Alfvén number* as defined by the ratio of the characteristic fluid velocity to the propagating velocity of the *Alfvén* wave, which is given by $\sqrt{B^2/\mu_m\rho}$, is called a magnetic Mach *number* by many writers in view of the partial analogy with Mach number.

magnetic Prandtl number $Pr_m = \nu/\nu_m = Re_m/Re$ represents the ratio of the diffusion of the fluid vortex to that of the magnetic field. The value of this dimensionless number is very small in metallic materials. For example, this number for mercury is about 10^{-7}. This means that the diffusion of the magnetic field is much faster than that of the vortex.

magnetic pressure number $P_m = B^2/\rho\mu_m V^2$ represents the ratio of the magnetic pressure to the dynamic pressure of the fluid.

The other dimensionless number The dimensionless number $Ha^2Re/Gr = \sigma V B^2/\rho g\beta\Delta T$ represents the ratio of the buoyancy force to the Lorenz force. Here, Gr is called Grashof number, which is defined as the ratio of (buoyancy force) × (inertial force) to (viscous force)^2 and given by $\rho^2 g\beta\Delta T L^3/\mu^2$. Thus, Ha^2/Gr expresses the ratio of (Lorenz force) × (viscous force) to (buoyant force) × (inertial force).

7 Solutions of Problems 3.1–3.12

The solution of Problem 3.1

$$\rho\{\partial \boldsymbol{v}/\partial t + (\boldsymbol{v}\cdot\nabla)\boldsymbol{v}\} = -\nabla \boldsymbol{p} + \mu\nabla^2\boldsymbol{v} + \boldsymbol{f} \tag{3.2}$$

By dividing both sides of Eq. 3.2 by ρ, Eq. 1 is obtained.

$$\partial \boldsymbol{v}/\partial t + (\boldsymbol{v}\cdot\nabla)\boldsymbol{v} = -\nabla p/\rho + (\mu/\rho)\nabla^2\boldsymbol{v} + \boldsymbol{f}/\rho \tag{1}$$

By operating the rotation, we get

$$\nabla\times\partial \boldsymbol{v}/\partial t + \nabla\times(\boldsymbol{v}\cdot\nabla)\boldsymbol{v} = -\nabla\times\nabla p/\rho + (\mu/\rho)\nabla\times\nabla^2\boldsymbol{v} + \nabla\times\boldsymbol{f}/\rho \tag{2}$$

By using the definition of vorticity $\nabla\times\boldsymbol{v}\equiv\boldsymbol{\omega}$ and the vector identity $\nabla\times\nabla\boldsymbol{A} = 0$ ($\boldsymbol{A}$ is an arbitrary vector), Eq. 3 is obtained as follows:

$$\begin{aligned}&\partial\boldsymbol{\omega}/\partial t + \nabla\times\nabla(\boldsymbol{v}\cdot\boldsymbol{v})/2 - \nabla\times(\boldsymbol{v}\times\boldsymbol{\omega})\\ &\quad= -\nabla\times\nabla p/\rho + (\mu/\rho)\nabla^2\boldsymbol{\omega} + \nabla\times\boldsymbol{f}/\rho\end{aligned} \tag{3}$$

Then, by assigning $\boldsymbol{f} = \boldsymbol{J}\times\boldsymbol{B}$ and $\mu/\rho = \nu$, Eq. 3.3 can be derived.

$$\partial\boldsymbol{\omega}/\partial t = \nabla\times(\boldsymbol{v}\times\boldsymbol{\omega}) + \nu\nabla^2\boldsymbol{\omega} + \nabla\times(\boldsymbol{J}\times\boldsymbol{B})/\rho \tag{3.3}$$

The solution of Problem 3.2

Equation 3.6 is expressed as follows:

$$\nabla\times\boldsymbol{H} = \boldsymbol{J} = \begin{vmatrix}\boldsymbol{i}_x & \boldsymbol{i}_y & \boldsymbol{i}_z\\ \partial/\partial x & \partial/\partial y & \partial/\partial z\\ H_x & H_y & H_z\end{vmatrix}$$

$$= \left(\frac{\partial H_z}{\partial y} - \frac{\partial H_y}{\partial z}\right)\boldsymbol{i}_x + \left(\frac{\partial H_x}{\partial z} - \frac{\partial H_z}{\partial x}\right)\boldsymbol{i}_y + \left(\frac{\partial H_y}{\partial x} - \frac{\partial H_x}{\partial y}\right)\boldsymbol{i}_z$$

Then, $\nabla\cdot\boldsymbol{J} = \nabla\cdot(\nabla\times\boldsymbol{H})$

$$= \partial\left(\frac{\partial H_z}{\partial y} - \frac{\partial H_y}{\partial z}\right)\Big/\partial x + \partial\left(\frac{\partial H_x}{\partial z} - \frac{\partial H_z}{\partial x}\right)\Big/\partial y + \partial\left(\frac{\partial H_y}{\partial x} - \frac{\partial H_x}{\partial y}\right)\Big/\partial z = 0 \tag{3.9}$$

The solution of Problem 3.3

By rewriting Eq.(3.11),

$$\boldsymbol{E} = (\boldsymbol{J}/\sigma) - (\boldsymbol{v}\times\boldsymbol{B}) \tag{1}$$

From Eqs. 3.6 and 3.10,

$$\boldsymbol{J} = (\nabla \times \boldsymbol{B})/\mu_m \tag{2}$$

By substituting Eq. 2 into Eq. 1,

$$\boldsymbol{E} = (\nabla \times \boldsymbol{B})/\sigma\mu_m - (\boldsymbol{v} \times \boldsymbol{B}) = \nu_m(\nabla \times \boldsymbol{B}) - (\boldsymbol{v} \times \boldsymbol{B}) \qquad \because \nu_m = 1/\sigma\mu_m \tag{3}$$

Equation 3.5 is rewritten as $-\partial \boldsymbol{B}/\partial t = \nabla \times \boldsymbol{E}$. Thus, from Eqs. 3.5 and 3,

$$\begin{aligned} -\partial \boldsymbol{B}/\partial t &= \nu_m \{\nabla \times (\nabla \times \boldsymbol{B})\} - \nabla \times (\boldsymbol{v} \times \boldsymbol{B}) \\ &= \nu_m \left\{\nabla(\nabla \cdot \boldsymbol{B}) - \nabla^2 \boldsymbol{B}\right\} - \nabla \times (\boldsymbol{v} \times \boldsymbol{B}) \end{aligned}$$

By using Eq. 3.7 of $\nabla \cdot \boldsymbol{B} = 0$

$$\because \partial \boldsymbol{B}/\partial t = \nabla \times (\boldsymbol{v} \times \boldsymbol{B}) + \nu_m \nabla^2 \boldsymbol{B} \tag{3.12}$$

The solution of Problem 3.4

$$\nabla \times \boldsymbol{H} = \left(\frac{\partial H_z}{\partial y} - \frac{\partial H_y}{\partial z}\right)\boldsymbol{i}_x + \left(\frac{\partial H_x}{\partial z} - \frac{\partial H_z}{\partial x}\right)\boldsymbol{i}_y + \left(\frac{\partial H_y}{\partial x} - \frac{\partial H_x}{\partial y}\right)\boldsymbol{i}_z = \boldsymbol{0} \tag{1}$$

Therefore

$$\left(\frac{\partial H_z}{\partial y} - \frac{\partial H_y}{\partial z}\right) = \left(\frac{\partial H_x}{\partial z} - \frac{\partial H_z}{\partial x}\right) = \left(\frac{\partial H_y}{\partial x} - \frac{\partial H_x}{\partial y}\right) = 0 \tag{2}$$

The other hand, if the magnetic field $\boldsymbol{H}$ is a conservative field, Eq. 3 should hold.

$$\boldsymbol{H} = -grad\,\phi \tag{3}$$

Here, ϕ is the scalar potential of a magnetic field. Equation 3 always is satisfied with the relation of Eq. 2 as shown in Eq. 4.

$$\frac{\partial}{\partial y}\frac{\partial \phi}{\partial z} - \frac{\partial}{\partial z}\frac{\partial \phi}{\partial y} = \frac{\partial}{\partial z}\frac{\partial \phi}{\partial x} - \frac{\partial}{\partial x}\frac{\partial \phi}{\partial z} = \frac{\partial}{\partial x}\frac{\partial \phi}{\partial y} - \frac{\partial}{\partial y}\frac{\partial \phi}{\partial x} = 0 \tag{4}$$

Therefore, it is understood that the scalar potential of a magnetic field ϕ exists and its function is arbitrary.

The solution of Problem 3.5

$$d^2 v_x/dy^2 - (\sigma/\mu) B_0{}^2 v_x = (\partial p/\partial x + \sigma B_0 E_z)/\mu \tag{3.41}$$

By using $Ha \equiv y_0 B_0 \sqrt{\sigma/\mu}$

$$d^2 v_x/dy^2 - (Ha/y_0)^2 v_x = (1/\mu)(\partial p/\partial x) + (H_a/y_0)\sqrt{\sigma/\mu}\ E_z \tag{1}$$

A general solution of Eq. 1 is given as follows:

$$v_x = \alpha\, exp\left(\frac{Ha}{y_0}y\right) + \beta\, exp\left(-\frac{Ha}{y_0}y\right) - \left(\frac{y_0}{Ha}\right)^2\left(\frac{1}{\mu}\frac{\partial p}{\partial x} + \frac{Ha}{y_0}\sqrt{\frac{\sigma}{\mu}}\ E_z\right) \tag{2}$$

Equation 3 is obtained from the boundary conditions.

$$\left.\begin{aligned} \alpha\ exp(Ha) + \beta\ exp(-Ha) - \left(\frac{y_0}{Ha}\right)^2\left(\frac{1}{\mu}\frac{\partial p}{\partial x} + \frac{Ha}{y_0}\sqrt{\frac{\sigma}{\mu}}\ E_z\right) = 0 \\ \alpha\ exp(-Ha) + \beta\ exp(Ha) - \left(\frac{y_0}{Ha}\right)^2\left(\frac{1}{\mu}\frac{\partial p}{\partial x} + \frac{Ha}{y_0}\sqrt{\frac{\sigma}{\mu}}\ E_z\right) = 0 \end{aligned}\right\} \tag{3}$$

From Eq. 3, the following relation is obtained:

$$(\alpha - \beta)\ exp(Ha) + (\beta - \alpha)\ exp\,(-Ha) = 0$$

Then, Eq. 4 is obtained since $Ha \neq 0,\ exp(Ha) > 0$ and $exp(-Ha) > 0$ are.

$$\alpha = \beta \tag{4}$$

By substituting Eq. 4 into Eq. 3, Eq. 5 is obtained.

$$\alpha\,\{exp(-Ha) + exp(Ha)\} - \left(\frac{y_0}{Ha}\right)^2\left(\frac{1}{\mu}\frac{\partial p}{\partial x} + \frac{Ha}{y_0}\sqrt{\frac{\sigma}{\mu}}\ E_z\right) = 0$$

$$\therefore \quad \alpha = \frac{1}{2\cosh Ha}\left(\frac{y_0}{Ha}\right)^2\left(\frac{1}{\mu}\frac{\partial p}{\partial x} + \frac{Ha}{y_0}\sqrt{\frac{\sigma}{\mu}}\ E_z\right) \tag{5}$$

Substituting Eq. 5 into Eq. 2 and arranging the resulting equation yields the following equation:

$$\begin{aligned} v_x &= \frac{1}{2\cosh Ha}\left(\frac{y_0}{Ha}\right)^2\left(\frac{1}{\mu}\frac{\partial p}{\partial x} + \frac{Ha}{y_0}\sqrt{\frac{\sigma}{\mu}}\ E_z\right) \\ &\quad \times \{\exp\,(Ha\ y/y_0) + exp(-Ha\ y/y_0)\} - \left(\frac{y_0}{Ha}\right)^2\left(\frac{1}{\mu}\frac{\partial p}{\partial x} + \frac{Ha}{y_0}\sqrt{\frac{\sigma}{\mu}}\ E_z\right) \\ &= \left(\frac{y_0}{Ha}\right)^2\left(\frac{1}{\mu}\frac{\partial p}{\partial x} + \frac{Ha}{y_0}\sqrt{\frac{\sigma}{\mu}}\ E_z\right)\left\{\frac{\cosh(Ha y/y_0)}{\cosh Ha} - 1\right\} \end{aligned} \tag{3.42}$$

The solution of Problem 3.6

Equation 3.42 is expressed as follows:

$$v_x = \frac{y_0^2}{\cosh Ha}\left(\frac{1}{\mu}\frac{\partial p}{\partial x} + \frac{Ha}{y_0}\sqrt{\frac{\sigma}{\mu}}\,E_z\right)\left\{\frac{\cosh\,(Hay/y_0) - \cosh Ha}{Ha^2}\right\} \tag{1}$$

By using L'Hopital's theorem, which provides the limit value of the function in which the numerator and denominator approach zero at the limit, the limit value is obtained as follows:

$$\begin{aligned}
&\lim_{Ha\to 0}\left\{\frac{\cosh\,(Hay/y_0) - \cosh Ha}{Ha^2}\right\}\\
&= \lim_{Ha\to 0}\left\{\frac{(y/y_0)\sinh\,(Hay/y_0) - \sinh Ha}{2Ha}\right\}\\
&= \lim_{Ha\to 0}\left\{\frac{(y/y_0)^2\cosh\,(Hay/y_0) - \cosh Ha}{2}\right\} = \frac{1}{2}\left(\frac{y^2}{y_0^2} - 1\right)
\end{aligned} \tag{2}$$

Thus, the following equation is obtained:

$$\begin{aligned}
\lim_{Ha\to 0} v_x &= \lim_{Ha\to 0}\frac{y_0^2}{\cosh\,Ha}\left(\frac{1}{\mu}\frac{\partial p}{\partial x}+\frac{Ha}{y_0}\sqrt{\frac{\sigma}{\mu}}\,E_z\right)\left\{\frac{\cosh\,(Hay/y_0) - \cosh Ha}{Ha^2}\right\}\\
&= \frac{y_0^2}{1}\cdot\frac{1}{\mu}\frac{\partial p}{\partial x}\cdot\frac{1}{2}\left(\frac{y^2}{y_0^2} - 1\right) = (1/2\mu)\,(\partial p/\partial x)\,y_0^2\left\{(y/y_0)^2 - 1\right\}
\end{aligned}$$

The solution of Problem 3.7

$$\begin{aligned}
\bar{v}_x &= \frac{1}{2y_0}\int_{-y_0}^{y_0} v_x dy\\
&= \frac{1}{2y_0}\frac{y_0^2}{Ha^2}\left(\frac{1}{\mu}\frac{\partial p}{\partial x} + \frac{Ha}{y_0}\sqrt{\frac{\sigma}{\mu}}\,E_z\right)\int_{-y_0}^{y_0}\left\{\frac{1}{\cosh Ha}\cosh\left(Ha\frac{y}{y_0}\right) - 1\right\} dy\\
&= \frac{1}{2y_0}\frac{y_0^2}{Ha^2}\left(\frac{1}{\mu}\frac{\partial p}{\partial x} + \frac{Ha}{y_0}\sqrt{\frac{\sigma}{\mu}}\,E_z\right)\left[\frac{y_0}{Ha\cosh Ha}\sinh\left(Ha\frac{y}{y_0}\right) - y\right]_{-y_0}^{y_0}\\
&= \frac{1}{2y_0}\frac{y_0^2}{Ha^2}\left(\frac{1}{\mu}\frac{\partial p}{\partial x} + \frac{Ha}{y_0}\sqrt{\frac{\sigma}{\mu}}\,E_z\right)\left(\frac{2y_0\sinh Ha}{Ha\cosh Ha} - 2y_0\right)\\
&= \frac{y_0^2}{Ha^2}\left(\frac{1}{\mu}\frac{\partial p}{\partial x} + \frac{Ha}{y_0}\sqrt{\frac{\sigma}{\mu}}E_z\right)\left(\frac{\sinh Ha}{Ha\cosh Ha} - 1\right)\\
&= \frac{y_0^2}{Ha^2}\left(\frac{1}{\mu}\frac{\partial p}{\partial x} + \frac{Ha}{y_0}\sqrt{\frac{\sigma}{\mu}}\,E_z\right)\left(\frac{\tanh Ha}{Ha} - 1\right)
\end{aligned} \tag{3.43}$$

The solution of Problem 3.8

$$v_x = \frac{y_0^2}{Ha^2}\left(\frac{1}{\mu}\frac{\partial p}{\partial x} + \frac{Ha}{y_0}\sqrt{\frac{\sigma}{\mu}}\,E_z\right)\left\{\frac{\cosh(Ha\;y/y_0)}{\cosh Ha} - 1\right\} \tag{3.42}$$

The amplitude of v_x is expressed by $\frac{y_0^2}{Ha^2}\left(\frac{1}{\mu}\frac{\partial p}{\partial x} + \frac{Ha}{y_0}\sqrt{\frac{\sigma}{\mu}}\,E_z\right)$
The limit of the amplitude is expressed as follows:

$$\begin{aligned}
&\lim_{\mathrm{Ha}\to\infty}\frac{y_0^2}{Ha^2}\left(\frac{1}{\mu}\frac{\partial p}{\partial x} + \frac{Ha}{y_0}\sqrt{\frac{\sigma}{\mu}}\,E_z\right)\\
&= \lim_{\mathrm{Ha}\to\infty}\left(\frac{y_0^2}{Ha^2}\frac{1}{\mu}\frac{\partial p}{\partial x} + \frac{y_0}{Ha}\sqrt{\frac{\sigma}{\mu}}\,E_z\right) = \lim_{\mathrm{Ha}\to\infty}\frac{y_0}{Ha}\sqrt{\frac{\sigma}{\mu}}\,E_z\\
&(\because\; Ha^2 \gg Ha,\;\; at\;\; Ha\; \to \infty)
\end{aligned} \tag{1}$$

Therefore

$$\lim_{\mathrm{Ha}\to\infty} v_x = \lim_{\mathrm{Ha}\to\infty}\frac{y_0}{Ha}\sqrt{\frac{\sigma}{\mu}}\,E_z = \lim_{\mathrm{Ha}\to\infty}\frac{E_z}{B_0} \propto 1/B_0 \tag{2}$$

When $E_z = 0$, $\lim_{\mathrm{Ha}\to\infty} v_x = \lim_{\mathrm{Ha}\to\infty}\frac{y_0^2}{\mathrm{Ha}^2}\frac{1}{\mu}\frac{\partial \mathrm{p}}{\partial \mathrm{x}} = \lim_{\mathrm{Ha}\to\infty}\frac{1}{\sigma B_0^2}\frac{\partial p}{\partial x} \propto 1/B_0^2$ (3)

The solution of Problem 3.9

$$B_x\;(z,\;t) = b_x(z)\;exp\,(i\omega t) \tag{3.59}$$

$$i\omega b_x = \nu_m(d^2 b_x/dz^2) \tag{3.60}$$

First, a characteristic equation of Eq. 3.60 is obtained as

$$\nu_m\lambda^2 - i\omega = 0 \quad \therefore\;\; \lambda = \pm(i\omega/\nu_m)^{1/2}$$

Here, by using the relation equation $\sqrt{i} = (i+1)/\sqrt{2}$, λ is expressed as follows:

$$\lambda = \pm(\omega/2\;\nu_m)^{1/2}\cdot(i+1)$$

Therefore, a general solution of b_x is given as follows:

$$\begin{aligned}
b_x = C_1 exp\left(\sqrt{\omega/2\nu_\mathrm{m}}\;z\right)\;\cdot\;exp\left(i\,\sqrt{\omega/2\nu_m}\;z\right)\\
+ C_2\,exp\left(-\sqrt{\omega/2\nu_\mathrm{m}}\;z\right)\;\cdot\;exp\left(-i\,\sqrt{\omega/2\nu_m}\;z\right),
\end{aligned}$$

where C_1 and C_2 are constants.

By using a boundary condition of Eq. 3.61, the constants are determined as

$$C_1 = 0, \quad C_2 = B_0$$

Then, $b_x = B_0\, exp(-\sqrt{\omega/2v_m}z) \cdot exp(-i\sqrt{\omega/2v_m}z)$

$$B_x(z, t) = b_x(z)\, exp(i\omega t)$$

$$B_x(z, t) = B_0 exp\left(-\sqrt{\omega/2v_m}\, z\right) \cdot exp\left\{i\left(-\sqrt{\omega/2v_m}\, z + \omega t\right)\right\} \tag{3.62}$$

The solution of Problem 3.10

$$\begin{aligned}\bar{f_z} &= \frac{1}{2} Re\left(-j_y \cdot b_x^*\right)\\ &= \frac{1}{2} Re\left[-\left(\frac{-1}{\mu_m}\right)\sqrt{\omega/2v_m}\,(1+i)B_0 exp\left(-\sqrt{\omega/2v_m}\, z\right) \cdot exp\, i\left(-\sqrt{\omega/2v_m}\, z\right)\right.\\ &\qquad \left.\times\, B_0 exp\left(-\sqrt{\omega/2v_m}\, z\right) \cdot exp\, \left\{-i\left(-\sqrt{\omega/2v_m}\, z\right)\right\}\right]\\ &= \left(B_0^2/2\delta\mu_m\right)\, exp(-2z/\delta)\end{aligned}$$

By integrating $\bar{f_z}$ over $0 \sim \infty$ in the z-direction, Eq. 3.71 is obtained.

$$\begin{aligned}P_m &= \int_0^\infty \bar{f_z} dz = \int_0^\infty \left(B_0^2/2\delta\mu_m\right)\, exp(-2z/\delta)\, dz\\ &= \frac{B_0^2}{2\delta\mu_m}\left(-\frac{\delta}{2}\right)\left[exp\left(-2\frac{z}{\delta}\right)\right]_0^\infty\\ &= B_0^2/4\mu_m = B_e^2/2\mu_m\end{aligned} \tag{3.71}$$

The solution of Problem 3.11

The average value of Ohmic dissipation per unit volume $\bar{q}$ is given in Eq. 1.

$$\bar{q} = (1/T)\int_0^T \left(J_y^2/\sigma\right) dt \tag{1}$$

By using Eq. 3.69, Eq. 1 is expressed by Eq. 2.

$$\bar{q} = (1/T)\int_0^T \left(J_y^2/\sigma\right) dt = Re\left(j_y^*\, j_y\right)/2\sigma \tag{2}$$

By taking account of the definition of j_y as $J_y = j_y\, exp\,(i\omega t)$, from Eq. 3.63, j_y and j_y^* are reduced to

$$\left.\begin{aligned} j_y &= -\frac{1}{\mu_m}\sqrt{\frac{\omega}{2\nu_m}}(1+i)\ B_0\,exp\left(-\sqrt{\frac{\omega}{2\nu_m}}\,z\right)\cdot\ exp\,i\left(-\sqrt{\frac{\omega}{2\nu_m}}\,z\right) \\ j_y^* &= -\frac{1}{\mu_m}\sqrt{\frac{\omega}{2\nu_m}}(1-i)\ B_0\,exp\left(-\sqrt{\frac{\omega}{2\nu_m}}\,z\right)\cdot\ exp\,i\left(\sqrt{\frac{\omega}{2\nu_m}}\,z\right) \end{aligned}\right\} \tag{3}$$

Substituting Eq. 3 into Eq. 2 yields Eq. 4

$$\bar{q} = (1/2\sigma)\cdot(1/\mu_m^2)\cdot(\omega/2\nu_m)\cdot 2B_0^2\,exp(-2\sqrt{\omega/2\nu_m}z) \tag{4}$$

Substituting $\nu_m = (1/\mu_m\sigma)$ into Eq. 4, the Eq. 3.72 is obtained.

$$\bar{q} = (\omega B_0^2/2\mu_m)\,exp(-2\sqrt{\omega/2\nu_m}z) \tag{3.72}$$

The solution of Problem 3.12
From Eq. 3.71, B_0^2 is expressed as the following equation.

$$B_0^2 = 4\mu_m p_m \tag{1}$$

Substituting B_0^2 into Eq. 3.73 yields Eq. 3.74.

$$Q = p_m\sqrt{2\omega/\mu_m\sigma} \tag{3.74}$$

References

1. W.F. Hughes, F.J. Young, *The Electromagnetodynamics of Fluids* (Wiley, New York, 1966), pp. 172, 174, 176
2. K. Ueno, Phys. Fluid. A **5**(2), 490 (1993)
3. T.C. Wang, S.J. Dudzinsk, AIAA J. **5**(1), 107 (1967)

Chapter 4
Electromagnetic Processing of Materials

In this chapter, we enumerate some functions of electric and magnetic fields as they act on electrically conductive fluids and we explain their contents on the basis of the knowledge of magnetohydrodynamics discussed in the previous chapter. Furthermore, based on these functions, we classify the technologies that use electromagnetic force. Then, we explore how to utilize each function in practical processes by giving some concrete examples. Since we believe that creation of new functions is the best way to fully understand the development of Electromagnetic Processing of Materials, we expect readers to actively propose and apply functions of their own devising.

1 Functions of Electric and Magnetic Fields Performing on Electrically Conductive Fluids

1.1 Shape Controlling Function

As shown in Fig. 4.1 a, a coordinate system is defined where the origin and the x-z plane are set on the surface of a molten metal, and the positive direction of the y-axis is taken outward on the basis of a right-handed coordinate system as shown in Fig. 4.1 b. By passing an electric current through a conducting wire, whose width is large enough compared with the metal surface area and which is set on the outside of the metal surface and parallel to the x-direction, a uniform magnetic field B_z is generated in the z-direction based on Eq. 3.6. A current density $\boldsymbol{J}$ induced in the metal by B_z can be calculated from Eq. 3.6 as follows:

$$J_x = (1/\mu_m)\,\partial B_z/\partial y \tag{4.1}$$

$$J_y = -(1/\mu_m)\,\partial B_z/\partial x \tag{4.2}$$

S. Asai, *Electromagnetic Processing of Materials*, Fluid Mechanics and Its Applications 99, DOI 10.1007/978-94-007-2645-1_4,
© Springer Science+Business Media B.V. 2012

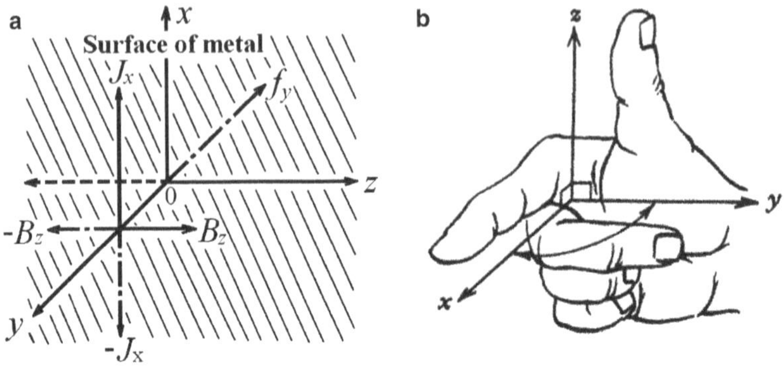

Fig. 4.1 **a** Schematic view for explaining the shape controlling function. **b** A right-handed coordinated system

Provided that B_z is uniformly distributed in the x-direction, $J_y = 0$ is obtained. The electromagnetic body force induced in the metal by the product of J_x and B_z can be calculated by substituting J_x and B_z into Eq. 3.17.

$$f_y = -J_x \cdot B_z = -(1/\mu_m)(\partial B_z/\partial y) \cdot B_z = -(1/2\mu_m)\left(\partial B_z{}^2/\partial y\right) \tag{4.3}$$

The pressure obtained by integrating the electromagnetic body force f_y, in the y-direction from $y = -\infty$ (which means a deeper position than electric skin depth) to $y = 0$ (surface) in a unit area is called the magnetic pressure p_m.

$$p_m = \int_{-\infty}^{0} f_y dy = -(1/2\,\mu_m)\,B_0{}^2 \tag{4.4}$$

Here, the boundary conditions are given as follows; at $y = 0$, $B_z = B_0$ and at $y = -\infty$, $B_z = 0$. As seen in Eq. 4.4, p_m has a negative value, which means that p_m works to push a molten metal from the surface to its inside. According to the theoretical derivation introduced so far, a static magnetic field induced by imposing a DC electric current in the x-direction could have generated a magnetic pressure. In such a case, however, p_m actually is not generated because the magnetic field diffuses and $(\partial B_z/\partial y) \to 0$ with time. On the other hand, in the case when an alternating magnetic field is induced by imposing an alternating current in the x-direction, p_m will not be zero because B_z always has its gradient in the y-direction. Moreover, since p_m obviously takes a negative value regardless of the electric current direction passing in the conducting wire, the molten metal will receive a compression force from the outside as shown in Eq. 4.4. Based on this principle, it is possible to control the shape of the molten metal by imposing an alternating current passing through a coil located at the outside of the metal surface.

The derivation of Eq. 4.4 is generalized as follows: By substituting Eq. 3.6 into Eq. 3.17 and using vector identities, Eq. 4.5 can be obtained.

$$\begin{aligned} \boldsymbol{f} &= \boldsymbol{J} \times \boldsymbol{B} = (1/\mu_m)(\nabla \times \boldsymbol{B}) \times \boldsymbol{B} \\ &= \underbrace{(1/\mu_m)(\boldsymbol{B} \cdot \nabla)\boldsymbol{B}}_{f_1} - \underbrace{\nabla(B^2/2\mu_m)}_{f_2} \end{aligned} \tag{4.5}$$

Equation 4.5 was already derived in the previous chapter as Eq. 3.18. When taking a rotation (which is equivalent to applying $\times$ in a vector operation) of $f_1 = (1/\mu_m)(\boldsymbol{B} \cdot \nabla)\boldsymbol{B}$ in Eq. 4.5, it does not necessarily take the value of zero $\{\nabla \times (\boldsymbol{B} \cdot \nabla)\boldsymbol{B}/\mu_m \neq 0\}$, but the rotation of $f_2 = \nabla(B^2/2\mu_m)$ should always be zero regardless of the distribution of $\boldsymbol{B}$ $\{\because \nabla \times (\nabla B^2/2\mu_m) = 0\}$. That is, f_2 must be a non-rotational force, while f_1 can possibly be a rotational force. As we can understand from Eq. 3.3, a fluid will be driven when $\boldsymbol{f}$ $(= \boldsymbol{J} \times \boldsymbol{B})$ is a rotational force, while it will not be driven when $\boldsymbol{f}$ is a non-rotational force, because the generation term of vorticity $\boldsymbol{\omega}$ does not exist. Therefore, when $\boldsymbol{f}$ is a non-rotational force in the whole area filled with the fluid, the shape of the fluid will be changed by a static pressure in the fluid so as to balance with $\boldsymbol{f}$. That is, only the non-rotational force acts on the fluid as a compression force that changes the fluid's shape.

Next, let us derive the ratio of f_1 to f_2 by using the equations given in the one-dimensional semi-infinite model in Sect. 4.2 of Chap. 3. f_1 is derived as shown in Eq. 4.6.

$$f_1 = (1/\mu_m)\,(\boldsymbol{B} \cdot \nabla)\,\boldsymbol{B} = (1/\mu_m)\,\{B_x(\partial/\partial \mathrm{x}) + B_y(\partial/\partial \mathrm{y}) + B_z(\partial/\partial \mathrm{z})\}\,\boldsymbol{B} \tag{4.6}$$

Because B_x is the only component of $\boldsymbol{B}$ and the other components do not exist in the case of the one-dimensional semi-infinite model, the value of f_1 can be approximated by Eq. 4.7.

$$f_1 \approx (1/\mu_m)\,B_x\,(\partial B_x/\partial x) \approx (1/\mu_m)\,(\partial {B_0}^2/L) \tag{4.7}$$

Although B_x does not distribute in the x-axis direction in the one-dimensional semi-infinite model, the distribution of B_x in the x-axis direction inevitably exists in practical processes. So by using the characteristic length L of an apparatus, the gradient of B_x in the x-direction has been approximated as the following equation:

$$\partial B_x/\partial x \approx B_0/L \tag{4.8}$$

On the other hand, B_x has the largest gradient in the z-direction as shown in Eq. 3.62. Thus f_2 becomes

$$f_2 = \nabla\,(B^2/2\mu_m) \approx \partial\,(B_x^2/2\mu_m)/\partial z \tag{4.9}$$

By substituting B_x given in Eq. 3.62 into Eq. 4.9, f_2 is expressed by the equation.

$$f_2 = (1/2\mu_m)\left\{-2\sqrt{\omega/2v_m}\,(1+i)\right\} B_0^2 = (1/\mu_m)\{-(1+i)/\delta\}\,B_0^2 \tag{4.10}$$

Fig. 4.2 A schematic view for explaining the function of flow suppressing

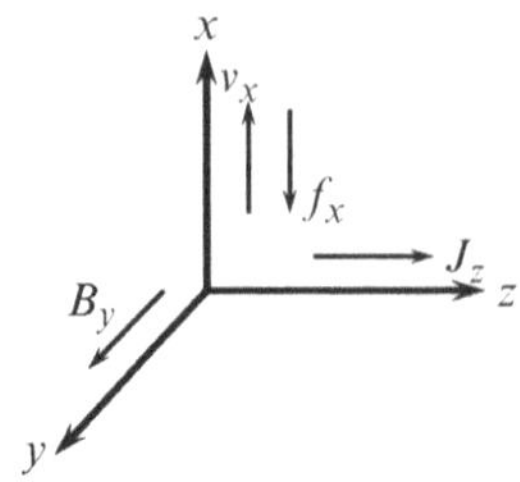

By using Eqs. 4.7 and 4.10, the ratio of f_2 to f_1 is expressed by the equation

$$|f_2/f_1| \approx L/\delta \tag{4.11}$$

Here, $\delta = (2/\mu_m \sigma \omega)^{1/2}$ is the skin depth defined by Eq. 3.64, and ω is the angular frequency in an alternating current.

When ω increases, the value of δ decreases and the value of $|f_2/f_1|$ increases. That is, the Lorentz force works entirely as a non-rotational force when ω is large. Therefore, an alternating magnetic field with high frequency is used for shape controlling of a molten metal, while a low frequency is used for driving (or mixing) of it, which will be explained in Sect. 1.5 in this chapter.

1.2 Flow Suppressing Function

When motion of a fluid v_x takes place in the x-direction under the imposition of the magnetic flux in the y-direction B_y as shown in Fig. 4.2, an electric current will be induced in the z-direction based on Eq. 3.11 as shown in Eq. 4.12.

$$\boldsymbol{J} = \sigma \boldsymbol{v} \times \boldsymbol{B} = (0, 0, \sigma v_x B_y) \tag{4.12}$$

If $J_z = \sigma v_x B_y$ is substituted into Eq. 3.17, it becomes

$$\boldsymbol{f} = \boldsymbol{J} \times \boldsymbol{B} = \left(-J_z B_y, 0, 0\right) = \left(-\sigma v_x B_y^2, 0, 0\right) \tag{4.13}$$

That is, a body force f_x of $-\sigma v_x B_y^2$ is generated in the x-direction. This body force f_x acts on the opposite direction of v_x due to its negative sign and implies that it is a suppression force against v_x. The same situation takes place when the motion of the fluid v_z exists in the z-direction. In this case the induced force will be $f_z = -\sigma v_z B_y^2$. That is, the motion of a fluid is suppressed in all directions except the y-direction, which corresponds to the direction of the imposed magnetic field. In addition, there is a case where the current of J_z shown in Eq. 4.12 cannot be generated [1], depending on electric boundary conditions. In such a case we should know that f_x is not generated.

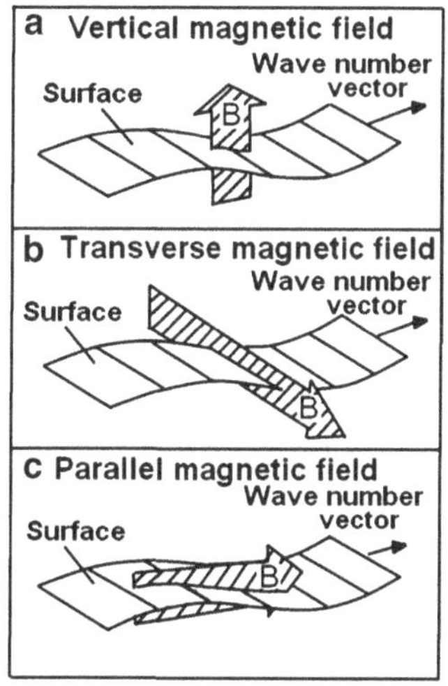

Fig. 4.3 Relation between the directions of imposing a magnetic field and propagating of wave motion

1.3 Wave Suppressing Function

When a molten metal is electrically conductive and a DC magnetic field is imposed, the wave motion is suppressed due to the same reason as the flow suppressing function mentioned in the previous Sect. 1.2. As shown in Fig. 4.3, there are three directions for imposing the DC magnetic field of **a** a vertical magnetic field, **b** a transverse magnetic field and **c** a parallel magnetic field, in the direction of the propagating wave. The details will be discussed in Sect. 2.3 of this chapter with the following results: **a** in a vertical magnetic field, the wave motion is attenuated in proportion to the square of the magnetic flux density, **b** in a transverse magnetic field, the flow suppressing function appears only when a magnetic field gradient exists along the liquid-depth direction, and **c** in a parallel magnetic field, the wave motion is attenuated nearly in proportion to the square of the magnetic flux density.

On the other hand, a high frequency AC magnetic field is considered to have the effect of suppressing a surface wave motion on a molten metal due to the pinch effect (magnetic pressure), which more strongly appears at a wave crest due to higher magnetic flux density, while it very weakly appears in a trough due to its lower magnetic flux density [2]. However, not enough theoretical and experimental researches have yet been done and a concrete theory does not exist at present.

1.4 Separating and Agglomerating Functions

As shown in Fig. 4.4, when a direct electric current and a static magnetic field, perpendicular to each other, are applied to a molten metal, the Lorentz force is hardly caused in an inclusion, because the electric current cannot flow into an inclusion that

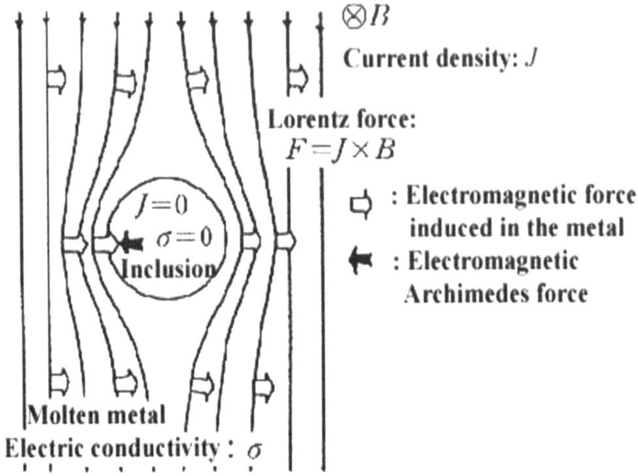

Fig. 4.4 Separating inclusions by use of electromagnetic force

has low electric conductivity, although the Lorentz force works in the molten metal. Therefore, the inclusion receives a counterforce from the metal surrounding it, and moves in the direction opposite to the direction of the Lorentz force caused in the molten metal. In this case, the force that the inclusion receives from the molten metal is called "**electromagnetic Archimedes force**" [3].

On the other hand, if a particle has a high electric conductivity, which is even larger than that of the molten metal, the electric current will converge on the particle. So the Lorentz force acting on the particle is larger than that on the molten metal and thus, the particle will move in the direction of the Lorentz force.

1.5 Driving (Mixing) Function

When a direct current with a density of J_x (x-direction) and a direct magnetic flux with a density of B_y (y-direction) are imposed perpendicular to each other, an electromagnetic body force f_z will be induced based on Eq. 3.17.

$$f_z = J_x B_y \tag{4.14}$$

A similar phenomenon will take place when an alternating current with a density of $J_x = \sqrt{2}J_e sin(\omega t)$ and an alternating magnetic flux $B_y = \sqrt{2}B_e sin(\omega t - \beta)$ are imposed perpendicular to each other. In this case the body force changes over a period so that the time average value of the force over the period is derived as follows:

$$\bar{f_z} = (1/T)\int_0^T \sqrt{2}J_e sin(\omega t)\cdot\sqrt{2}B_e sin(\omega t - \beta)dt = J_e B_e \cos\beta \tag{4.15}$$

Here, β is the phase difference between the current and the magnetic field, T the period and J_e and B_e the effective values of the alternating current and magnetic field, respectively.

When both an alternating current and a static magnetic field, or a direct current and an alternating magnetic field, are imposed, a body force with its direction and amplitude changing with time will be induced. In the case when the periodical time of the alternating electric current or magnetic field is shorter than a mechanical relaxation time ($t_{rel} = \rho L^2/2\pi\mu$), the averaged body force over the cycle will work on the fluid, because the fluid motion cannot follow the periodic change of the body force. On the contrary, in the case when the periodical time of the alternating electric current or magnetic field is longer than the mechanical relaxation time, a switching motion will take place because the fluid can follow the periodic change of the body force.

On the other hand, to drive a molten metal without imposition of electric current, an AC traveling magnetic field should be used. The derivation of the electromagnetic body force was shown in Eqs. 3.102 and 3.107.

$$\bar{\boldsymbol{f}} = \begin{cases} \bar{f}_x = \dfrac{b_0^2\omega\sigma\gamma}{2\sqrt{\gamma^4+(\mu_m\omega\sigma)^2}} \cdot \dfrac{\cosh(2\beta_r z) - \cos(2\beta_i z)}{\cosh(2\beta_r z_0) + \cos(2\beta_i z_0)} \\ \bar{f}_y = 0 \\ \bar{f}_z = \dfrac{b_0^2\omega\sigma\gamma}{2\sqrt{\gamma^4+(\mu_m\omega\sigma)^2}} \cdot \dfrac{-\beta_j \sinh(2\beta_r z) + \beta_r sin(2\beta_i z)}{\cosh(2\beta_r z_0) + \cos(2\beta_i z_0)} \end{cases} \tag{3.102}$$

$$\bar{\boldsymbol{f}} = \begin{cases} \bar{f}_x = \dfrac{b_0^2\omega\sigma\gamma}{2\sqrt{\gamma^4+(\mu_m\omega\sigma)^2}} \cdot \dfrac{\cosh(2\beta_r z) + \cos(2\beta_i z)}{\cosh(2\beta_r z_0) - \cos(2\beta_i z_0)} \\ \bar{f}_y = 0 \\ \bar{f}_z = -\dfrac{b_0^2\omega\sigma}{2\sqrt{\gamma^4+(\mu_m\omega\sigma)^2}} \cdot \dfrac{\beta_i \sinh(2\beta_r z) + \beta_r sin(2\beta_i z)}{\cosh(2\beta_r z_0) - \cos(2\beta_i z_0)} \end{cases} \tag{3.107}$$

Here, in the relation of $\beta^2 = \gamma^2 + i\omega/\nu_m$, β_r, and β_i are the real and imaginary parts of β, respectively. The y-component of $(\nabla \times \boldsymbol{f})$ which is the rotation of $\boldsymbol{f}$ shown in Eqs. 3.102 and 3.107 is given by $(\nabla \times \bar{\boldsymbol{f}})_y = (\partial \bar{f}_x/\partial z - \partial \bar{f}_z/\partial x) = (\partial \bar{f}_x/\partial z - 0)$. We can understand that $\bar{f}_z$ does not contribute to the rotation. That is, when an alternating magnetic field traveling in the y-direction is imposed, the force in the z-direction $\bar{f}_z$ acts as the compression force that takes the shape controlling function and the force $\bar{f}_x$ in the x-direction contributes as the driving force that causes the vorticity in a fluid. In practical electromagnetic mixing devices, b_0 distributes in the x-direction so that $\bar{f}_z$ will also contribute to the rotation a little.

1.6 Oscillating (Vibrating) Function

Under a combined imposition of alternating current and static magnetic field or that of a direct current and an alternating magnetic field, oscillation is generated in a fluid when the induced electromagnetic body force is large enough to overcome the viscous force, even though the periodical time of the alternating electric current or magnetic field is shorter than a mechanical relaxation time of the fluid. When the oscillating frequency is high, an electromagnetic ultrasonic wave will be generated. When only a high frequency magnetic field is imposed, the electromagnetic ultrasonic wave could also be generated by the interaction of the magnetic field with an induced electric current [4].

1.7 Splashing Function

It is possible to make splashes of molten metal by imposing an electromagnetic body force larger than the gravity force or an adhesive force due to surface tension.

$$|\boldsymbol{J} \times \boldsymbol{B}| > \max \left\{|\rho \boldsymbol{g}|, 3\sigma_f / 2a^2\right\}, \tag{4.16}$$

where σ_f and a are the surface tension and the radius of curvature of a molten metal, respectively. Equation 4.16 expresses a splashing function.

1.8 Levitating (Gravity Changing) Function

By balancing the electromagnetic body force expressed by Eq. 3.17 and the gravity force, a material can be levitated. The balance condition is given by the following equation.

$$\boldsymbol{J} \times \boldsymbol{B} = \rho \boldsymbol{g} \tag{4.17}$$

The condition for levitating an object with the volume of V ($\equiv \int dv$) can be obtained by integrating Eq. 4.17.

$$\int \boldsymbol{J} \times \boldsymbol{B} dv = \boldsymbol{g} \int \rho dv \tag{4.17'}$$

The Lorentz force distributes in a material levitated by the high frequency magnetic field generated by a coil or cold crucible. In this case, a levitating condition is given as Eq. 4.17′. Thus, Eqs. 4.17 and 4.17′ express a levitating function.

Moreover, by imposing direct electric current and direct magnetic flux, it is possible to change the apparent acceleration coefficient of the gravity from $\boldsymbol{g}$ to $\boldsymbol{g}'$.

$$\rho \boldsymbol{g}' = \rho \left| \boldsymbol{g} - \boldsymbol{J} \times \boldsymbol{B}/\rho \right| \tag{4.18}$$

Equation 4.18 expresses a gravity changing function.

1.9 Temperature Raising Function

When a current $\boldsymbol{J}$ flows in an electrically conductive fluid, Joule heat is generated as shown in the following equation.

$$q = |\boldsymbol{J}|^2/\sigma \tag{4.19}$$

Equation 4.19 indicates a temperature raising function. Here, the current $\boldsymbol{J}$ is expressed based on Eqs. 3.5 and 3.11, and its origin is composed of three factors as follows:

$$\boldsymbol{J} = \underbrace{\boldsymbol{J}_0}_{\text{first term}} + \underbrace{\sigma(-\partial \boldsymbol{A}/\partial t)}_{\text{second term}} + \underbrace{\sigma \boldsymbol{v} \times \boldsymbol{B}}_{\text{third term}}, \tag{4.20}$$

where $\boldsymbol{A}$ is the vector potential of a magnetic flux (which is defined as $\boldsymbol{B} = \nabla \times \boldsymbol{A}, \nabla \cdot \boldsymbol{A} = 0$).

1. The first term is a current applied externally.
2. The second term is a current induced by an alternating magnet field
3. The third term is a current induced by the motion of an electrically conductive fluid under a magnetic field.

On the other hand, Joule heat generated by an alternating magnetic field is given by Eq. 3.72.

$$\bar{q} = \left(\omega B_0^2/2\mu_m\right) exp\left(-2\sqrt{\omega/2\nu_m}\, z\right) \tag{3.72}$$

This equation shows that the heat generation rate increases with increase of frequency ω and concentrates in a surface layer. Equation 3.72 expresses a temperature raising function by an alternating magnetic field.

1.10 Flow Velocity Detecting Function

When the electrically conductive fluid moves with flow velocity of $\boldsymbol{v}$ under the existence of a magnetic flux density $\boldsymbol{B}$, an electromotive force will be generated. This is called Fleming's right-hand rule as given by Eq. 4.21.

$$\boldsymbol{E} = -\boldsymbol{v} \times \boldsymbol{B} \tag{4.21}$$

Here, $\boldsymbol{v}$ can be obtained by measuring $\boldsymbol{E}$ based on Eq. 4.21 under the given value of $\boldsymbol{B}$.

1.11 Compound Function

A compound function is formed by combining several of the basic functions. A refining function is formed by combining functions of separating, driving, oscillating, splashing, levitating and temperature raising. A solidification structure controlling function is revealed by combining functions of flow controlling, driving, oscillating and levitating.

2 Processes Using Electromagnetic Force

In Table 4.1, the major processes are classified by the functions that electric and magnetic fields exert on an electrically conductive fluid. Here, each process will be explained by corresponding functions.

2.1 Shape Controlling Function

2.1.1 Electromagnetic Casting

In 1966, Electromagnetic Casting (EMC) was invented by Gezelev [5] in the Soviet Union. After that, aluminum manufacturing companies in Europe and the United States made efforts to develop this process, which now operates in a commercial setting [6]. Instead of using a mold in a continuous casting machine for solidification of molten metal, one needed only several turns of coils as shown in Fig. 4.5.

When these coils are powered on with a high frequency electrical current of one to several kHz, a high frequency magnetic field will be generated on the surface of the molten metal, and an electromagnetic force will be induced inside the molten metal. In this case, due to the high frequency, the skin depth δ is around 1 mm or less, and the electromagnetic force, namely a magnetic pressure $p_m = {B_e}^2/2\mu_m$ (refer to Eq. 3.71 for the derivation) will act on the surface of the molten metal. By balancing the magnetic pressure p_m and the pressure due to the surface tension $p_c = \sigma_f/r$, with the static pressure of a molten metal $p_s(=\rho g h)$ and the dynamic pressure due to the molten metal motion $p_{dyn}\left(=\rho v^2/2\right)$, the molten metal can be levitated without being in contact with a mold.

$$p_m + p_c = p_s + p_{dyn} \tag{4.22}$$

2.1.2 Soft Contacting Solidification

Though the EMC developed in the aluminum industry is a very attractive technology for continuous casting of molten steels, a large molten metal height of h has to be

Table 4.1 Classification of processes making use of electromagnetic force through the functions

Basic principles	Function for materials processing	Process
Lorentz force $\boldsymbol{f} = \boldsymbol{J} \times \boldsymbol{B}$	(a) Shape controlling $p_m = -B^2/2\mu_m$	Cold crucible, Levitation melting, Electromagnetic casting, Electromagnetic plastic deformation, Soft contacting solidification
	(b) Flow suppressing $\boldsymbol{f} = \sigma(\boldsymbol{v} \times \boldsymbol{B}) \times \boldsymbol{B}$	Magnetic Czochralski method, Electromagnetic brake, Electromagnetic dam
	(c) Wave suppressing $p_m = -B^2/2\mu_m$ $\boldsymbol{f} = \sigma(\boldsymbol{v} \times \boldsymbol{B}) \times \boldsymbol{B}$	Electromagnetic dam, Electromagnetic casting
	(d) Separating and agglomerating	Electromagnetic removal of inclusions
	(e) Driving (Mixing)	Mixing by DC electric current and DC magnetic field, Mixing by traveling magnetic field, Electromagnetic pump, Inclusions removal by centrifugal force
	(f) Oscillating (Vibrating)	Electromagnetic oscillation, Electromagnetic ultrasonic wave
	(g) Splashing $\lvert\boldsymbol{J} \times \boldsymbol{B}\rvert >$ $\max\left\{\lvert\rho \boldsymbol{g}\rvert, 3\sigma_f/2a^2\right\}$	Electromagnetic atomization
	(h) Levitating $\boldsymbol{J} \times \boldsymbol{B} = \rho \boldsymbol{g}$	Horizontal electromagnetic casting, Frequency controlling of bubbling
Joule heat	(i) Temperature raising $q = \lvert\boldsymbol{J}\rvert^2/2\sigma$	Cold crucible, Levitation melting, High frequency magnetic heating, Ohmic heating
Fleming's right-hand rule	(j) Flow velocity detecting $\boldsymbol{E} = -\boldsymbol{v} \times \boldsymbol{B}$	Velocity sensor
Compound function	(k) Refining (d) + (e) + (f) + (g) + (h) + (i)	Electromagnetic refining, Inclusions removal
	(l) Solidification structure controlling (b) + (e) + (f) + (h)	Crystal grain refining, Crystal grain coarsening, Single crystal growth, Super cooling solidification

held up because the casting velocity is over ten times faster than that of aluminum. In addition, the density ρ of iron is twice larger than that of aluminum so that it is difficult to satisfy the condition of $\rho gh = B_e^2/2\mu_m$, which is given as the fundamental principle of EMC. Thus, the practical use of EMC for molten steels has been considered to be unrealistic so far. Meanwhile, the concept of soft contacting

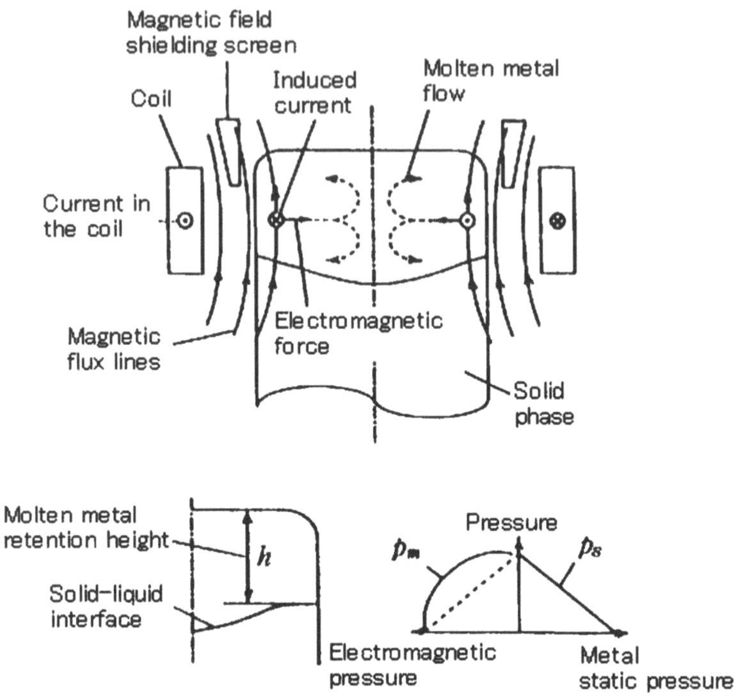

Fig. 4.5 Principle of electromagnetic casting

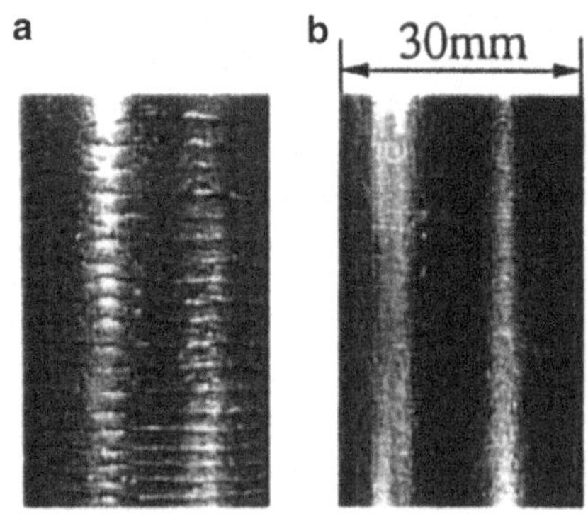

Fig. 4.6 The surface of tin billets
a without magnetic field,
b with magnetic field

solidification has been proposed where a high frequency magnetic field is imposed from the outside of a mold to improve surface quality of cast metals [7]. The surface aspect of the tin billets cast with and without a magnetic field is shown in Fig. 4.6. The surface quality of the sample cast with a magnetic field was remarkably improved in comparison with the sample treated without a magnetic field [8].

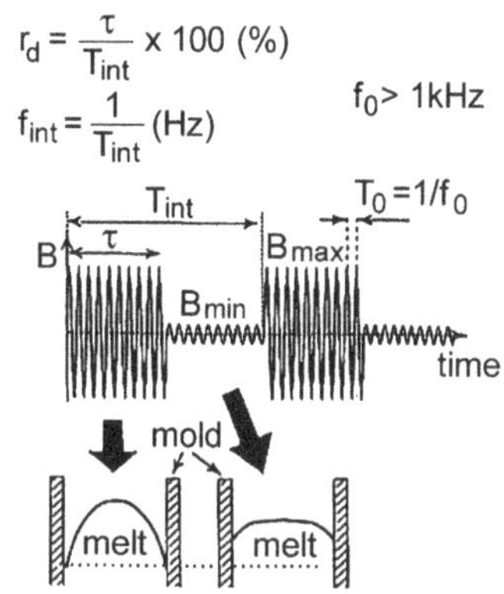

Fig. 4.7 Wave pattern of an intermittent high frequency magnetic field and shape variation of molten metal

In addition, a new method of imposing an intermittent high frequency magnetic field instead of the continuous high frequency magnetic field used for EMC has been proposed [9]. The variation of the intermittent high frequency magnetic field and the shape of molten metal are schematically shown in Fig. 4.7. The magnetic field consists of a high-frequency magnetic field with frequency over 1 kHz. The maximum and minimum amplitudes of the magnetic flux density B_{max} and B_{min} can be switched in a period of T_{int} between 1 and 1/100 s. This switching cycle is called an intermittent frequency ($f_{int} = 1/T_{int}$). Also it is possible to optionally set a ratio r_d in which B_{max} is imposing in one cycle in the intermittent frequency. The change in surface shape of molten metal periodically repeats during imposition of an intermittent magnetic field. In order to find the difference in surface aspects of tin billets cast under imposition of an intermittent high-frequency magnetic field or conventional high-frequency magnetic field (which will be referred to a continuous magnetic field hereafter), continuous casting of a molten tin was conducted under these two magnetic fields. The surface roughness of the cast metals is shown in Fig. 4.8. In this figure, symbols of □ and ○ indicate the experimental results obtained by imposing the continuous and intermittent magnetic fields, respectively. Under both of the two conditions, the surface roughness decreased with increase of coil current. To get a surface roughness indicated by the dotted line in the figure, the required electric current was drastically decreased from 400 to 120 A under an intermittent magnetic field.

It was reported that roughly half of the continuously cast steel slabs were once cooled to room temperature for surface treatment, and then reheated again for rolling. The application of EMC technology to casting of molten steel makes a direct rolling possible where the casts of steels are directly sent to a rolling process without surface treatment. According to the total amount of steel production in 1998 in Japan, the energy saved by direct rolling was estimated to be about 0.2% of the energy consumption in the whole country, which corresponds to the total energy consumption in a small prefecture in Japan [10].

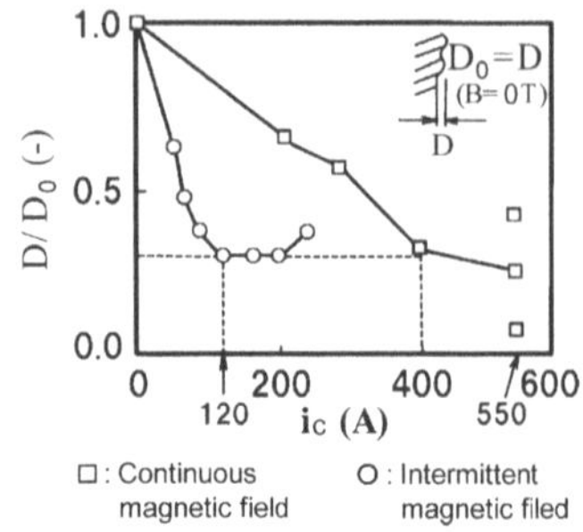

Fig. 4.8 The effect of electric current on the surface aspect of continuously cast tin

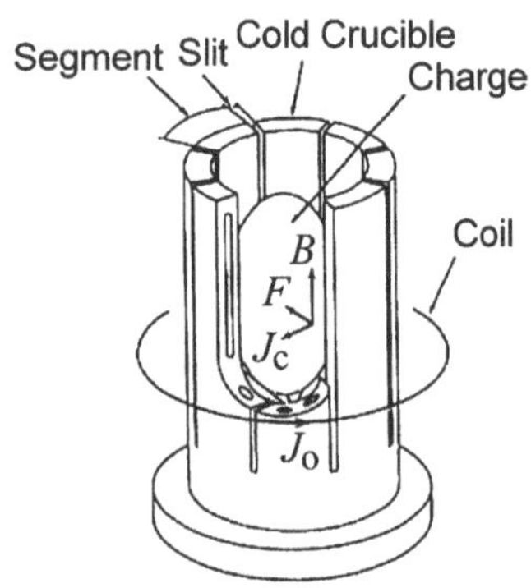

Fig. 4.9 A schematic view of a cold crucible

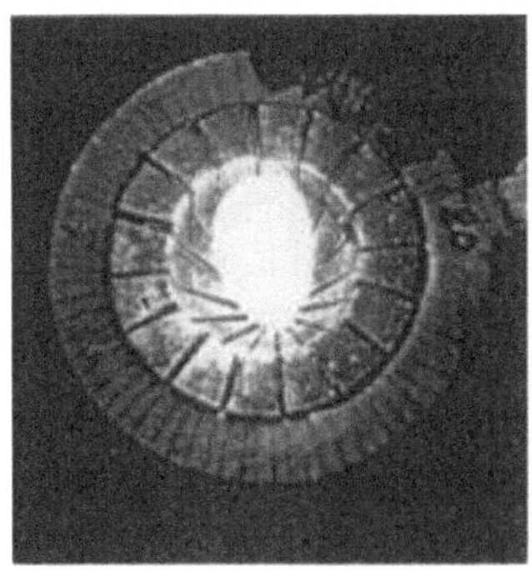

Fig. 4.10 Melting and holding of aluminum in a cold crucible

2.1.3 Cold Crucible

A cold crucible is an apparatus in which a copper crucible made of water-cooled segments is set inside a high-frequency induction coil (A schematic view is shown in Fig. 4.9). This crucible has the functions of no-contacting holding by use of magnetic pressure and melting of metals (see Fig. 4.10).

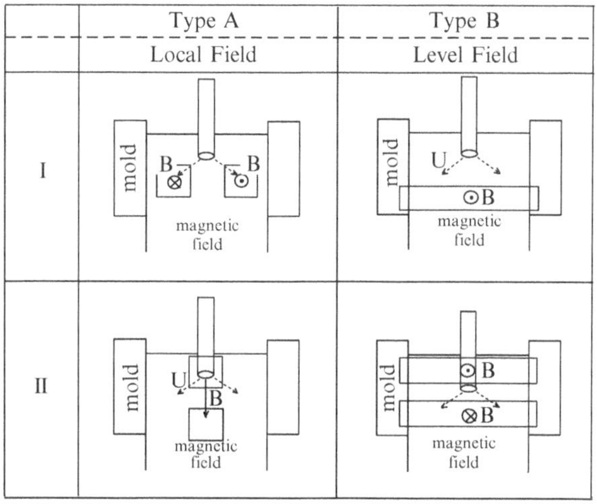

Fig. 4.11 Classification of methods imposing an electromagnetic brake [11]

2.2 *Flow Suppressing Function*

2.2.1 Electromagnetic Brake

The conventional electromagnetic brake in a continuous casting of steel belongs to type A in Fig. 4.11, where a magnetic field is imposed only around the exit of a nozzle. It has been modified to type B in the figure, where a magnetic field is imposed over the whole width of the cast slab. In addition, the type B can be classified into those with one or two stages. To answer the question of which one is superior, the one-stage imposition or the two-stage imposition, we need to consider not only the magnetohydrodynamic aspects but also the metallurgical.

The magnetic Czochralski method has been applied to the single crystal growth of Si, where the generation of turbulence is suppressed by imposing static magnetic field. This is a typical example making use of the electromagnetic brake [12].

2.2.2 Clad Slab

When a static magnetic field is so imposed as to crosscut the center of a continuous casting pool as shown in Fig. 4.12, the mixing between the upper and lower parts of molten steel will be suppressed.

By pouring a stainless steel in the upper region at flow rate Q_A and a plain steel in the lower region at flow rate Q_B, a clad steel having the stainless steel in its surface layer and the plain steel in its center part can be produced [11]. Figure 4.13 shows the difference between the dispersions of phosphorous tracer with and without imposing a static magnetic field. The effect of the imposed static magnetic field on the suppression of molten steel convection is clearly seen. This technology is

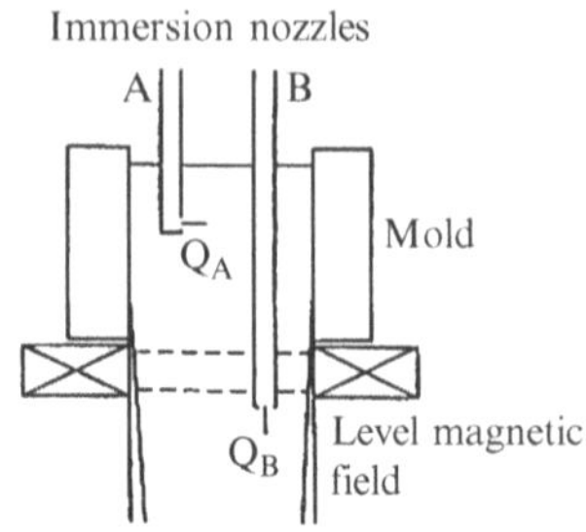

Fig. 4.12 Principle of clad slab casting by imposition of a magnetic field [11]

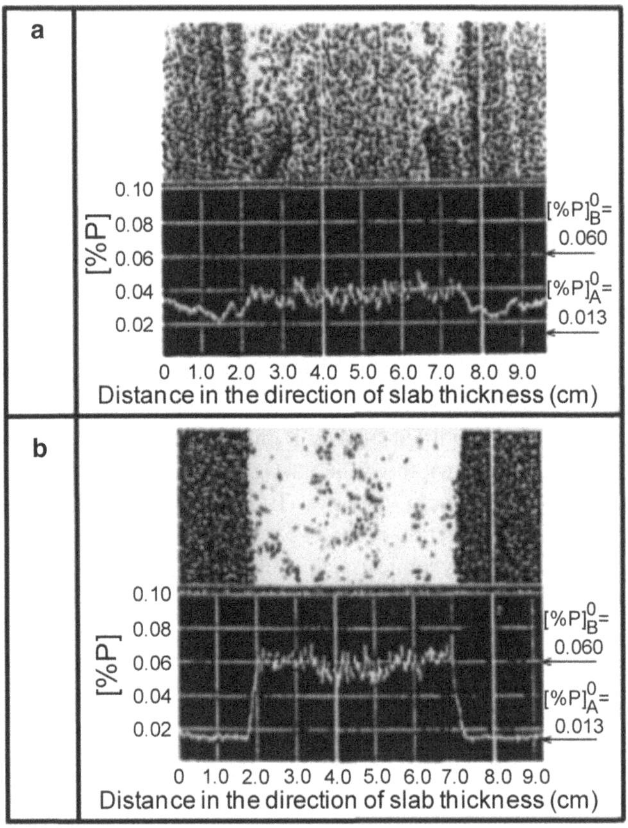

Fig. 4.13 Distribution of phosphorus in the section of a clad slab [11]. **a** without magnetic field, **b** with magnetic field (0.5 T)

highly promising since it not only allows mass production of clad steels, but also provides peculiar functions like producing of ultra thin stainless steel plates beyond the rolling limit.

2.3 Wave Suppressing Function

Shercliff [13] studied the suppression of a wave in the case of applying a DC magnetic field to the electrically conductive fluid, and Robinson [14] elucidated it both theoretically and experimentally. By considering the interface between two fluids having different densities from each other, Chandrasekhar [15] systematically summarized the effect of vertical and horizontal magnetic fields on the surface wave motion as an unstable phenomenon. Garnier and Moreau [2] theoretically indicated that the surface wave motion can be suppressed by using magnetic pressure induced by an alternating magnetic field.

On the other hand, researches on suppression of the surface wave motion to which attention has been paid especially in metallurgical processes, have been intensively studied. Kozuka et al. theoretically and experimentally quantified the suppression effects of the surface wave motion by the imposition of a static magnetic field in vertical [16], transverse [1] and horizontal [17] directions.

2.4 Separating and Agglomerating Functions

As shown in Fig. 4.14, there are four methods to induce the electromagnetic force in a molten metal. The method where electrodes are indispensable for a current to pass through them can be classified into two kinds: the direct imposition of a current and a magnetic field, and the imposition of only a current while a magnetic field is induced by the imposed current. On the other hand, for the methods where no electrodes are used, there exists the alternating magnetic field induced by a solenoid and the traveling magnetic field.

In 1954, Leenov and Kolin [18] clarified that spherical inclusions are forced under a uniform electromagnetic force. After that, Bepme [19] indicated in an experiment that inclusions could be moved by imposition of the electromagnetic force. Furthermore, in the 1980s, a theoretical formula for the separating and removal force that acts on an inclusion with cylindrical shape was given by Marty and Alemany [20]. On eliminating inclusions by an electromagnetic force, because molten metal flow is caused by the uneven distribution of electromagnetic force due to the gradient of a magnetic field, suppression of the flow is indispensable. Park et al. [21] proposed a method where an electromagnetic force is imposed on a molten metal flowing in channels that consist of many narrow tubes. Since the flow is suppressed in the narrow tube, the transfer, namely the separation of inclusions caused by the electromagnetic force, can be clearly observed.

On the other hand, the effect of the frequency was theoretically examined by Patel and El-Kaddah [22]. There are various methods for imposing an alternating electromagnetic force. Korovin [23] derived the theoretical formula on the imposition of the fixed alternating magnetic field. Taniguchi et al. [24] analyzed the process from the viewpoint of Chemical Reaction Engineering and found useful knowledge

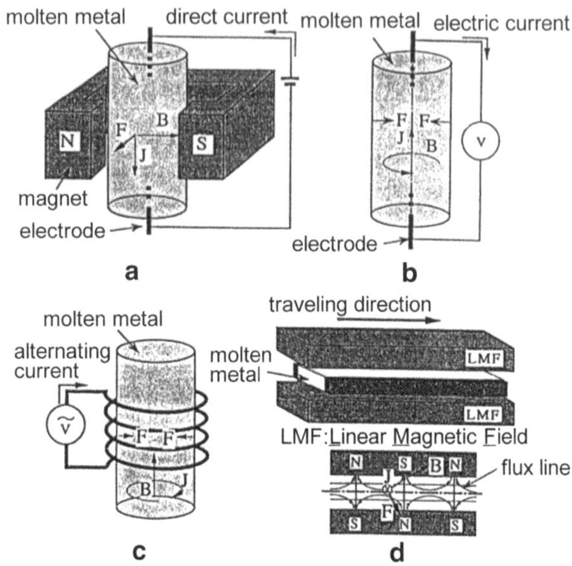

Fig. 4.14 Classification of electromagnetic removal of inclusions. **a** Imposition of direct (alternating) electric current and magnetic field, requirement of electrode. **b** Imposition of direct or alternating electric current, requirement of electrode. **c** Imposition of alternating magnetic field, no requirement of electrode. **d** Imposition of traveling magnetic field, no requirement of electrode

for practical operations. In addition, Tanaka et al. [25] proposed a method using a traveling magnetic field with commercial frequency and found a new function for the separation of inclusions by the traveling magnetic field. Yamao et al. [26] studied the removal of inclusions by imposition of a fixed alternating magnetic field from both experimental and theoretical aspects and clarified effects of the operation factors on their removal efficiency.

2.5 *Driving (Mixing) Function*

A driving function in the electromagnetic processing of materials has been applied to various manufacturing processes mainly for mixing of molten metals. Especially, electromagnetic mixing in the continuous casting process of steel is indispensable. Also, to satisfy the demand for high grade materials, the electromagnetic mixing processes have been studied for the purpose of agglomeration of inclusions in secondary refining processes. For other applications, an electromagnetic pump for pouring molten metals is a representative process using the driving function.

2.6 *Oscillating (Vibrating) Function*

By simultaneously imposing a static magnetic field of 0.7 T and an alternating current of 360 A with 60 Hz on a molten aluminum alloy (A356), Vivés et al.

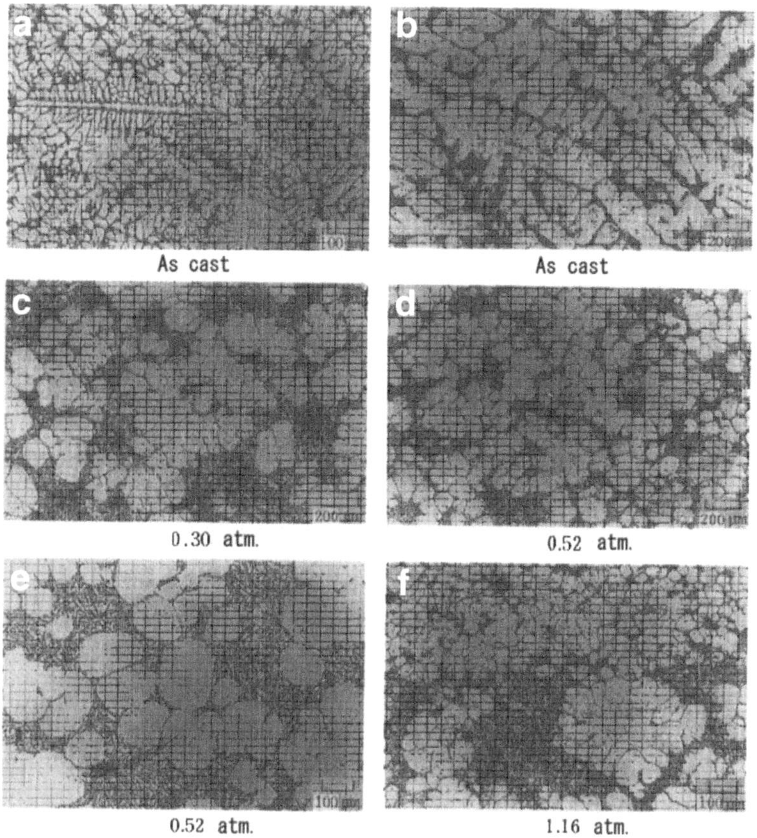

Fig. 4.15 Macro structure obtained by imposing electromagnetic vibration [27]

[27, 28] studied its effect on the solidified structure. The results are shown in Fig. 4.15. The solidification structures **a** and **b** were observed for the cases cast without imposing electro-magnetic force and rapidly or slowly cooled, respectively. Columnar dendrite structures can be seen in both cases. In the structure **c** on which an electromagnetic pressure of 0.30 atm was imposed, coarse dendrite flakes and large clusters exist. Both the structures **d** and **e** are obtained under the imposition of an electromagnetic pressure of 0.52 atm, where crystalline structures were made finer by electromagnetic vibration. The structure **f** is obtained under imposition of the electromagnetic pressure of 1.16 atm, which is strong enough to cause cavitations. In this case, crystals are finer than that without cavitation effect {**d** and **e**}. It can be observed that crystal structures were destroyed by cavitations generated by such a strong electromagnetic vibration.

Hitherto, it has been known that changes in macro and micro structures, refinement of crystal grains and degassing can be done by imposing sound and ultrasonic waves on metals during solidification [29]. The sound and ultrasonic

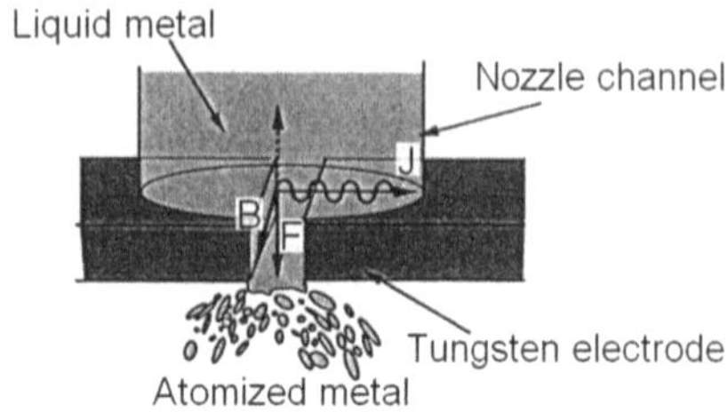

Fig. 4.16 A method for grain refinement of molten metal by simultaneous imposition of an alternating current and a DC magnetic field [30]

waves are generated by use of a magnetostriction or an electrostriction oscillator. Such oscillators, when they are made of quartz-crystal or graphite, are rapidly dissolved when they are dipped in a molten metal. The oscillator is also required to have enough strength because cavitations take place near the surface of the oscillator. From the viewpoint of cost and pollution of molten metal, it is difficult to apply the ultrasonic wave to processes treating a large amount of metal such as continuous casting processes and so on. Based on the consideration mentioned above, a non-contacting method to introduce an electromagnetic ultrasonic wave in a molten metal has been proposed by Amano et al. [4]. They detected a sound pressure having double frequency and double power intensity of the imposed magnetic field in an experiment using molten gallium. This result confirmed that an electromagnetic ultrasonic wave was generated by imposition of a high magnetic field.

2.7 *Splashing Function*

Varma and El-Kaddah [30] proposed a splashing process of molten metal, which makes use of the Lorentz force periodically generated by the simultaneous imposition of an alternating current and a direct magnetic field. Figure 4.16 shows the experimental apparatus. It has been reported that the atomization speed depends not on the intensity of the magnetic field but on the distance between electrodes. On the other hand, particle size distributions cannot be controlled by varying the intensity of the magnetic field and the distance between electrodes while it is possible to reduce particle diameters by intensifying the magnetic field.

Alemany et al. [31] proposed a grain refining method as shown in Fig. 4.17. A strong centrifugal force, which is induced by a rotating magnetic field imposed around a refractory nozzle, acts on a molten metal flow through the nozzle so that the molten metal is ejected from the nozzle to become a thin film and subsequently dispersed into tiny liquid drops due to surface tension.

On the other hand, Kolesnichenko et al. [32] has also proposed **a** an induction type and **b** a combination type of a magnetic field and electric current imposed from outside as shown in Fig. 4.18.

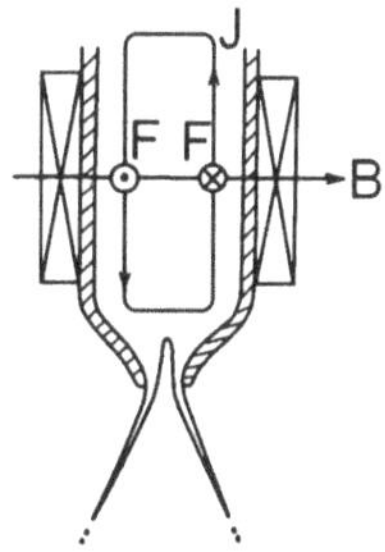

Fig. 4.17 Pulverizing method of molten metal by centrifugal force [31]

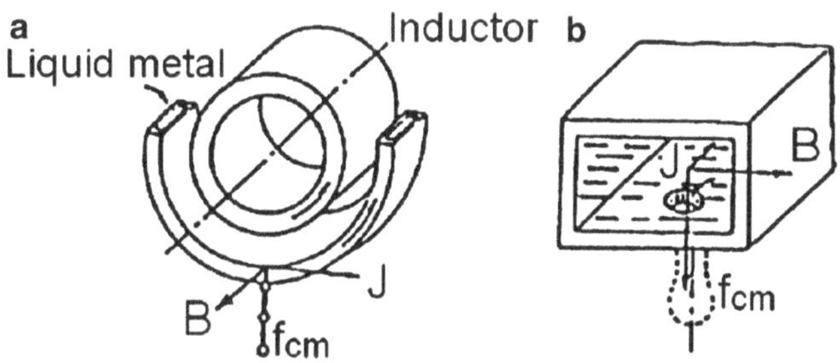

Fig. 4.18 Pulverizing method of molten metal by **a** an induction type and **b** a combination type of a magnetic field and electric current imposed from outside [32]

2.8 Temperature Raising Function

The electrical conductivities of such inorganic materials as ceramics, salts and glasses remarkably increase after melt-down, while they show poor electric conductivities in temperatures below their melting point. Therefore, in the melting operation of inorganic materials by using induction heating under the imposition of a high-frequency magnetic field, the melting can easily proceed by first introducing some amount of the melt, such as in the case of electrically conductive materials as metals. On the other hand, in the vicinity of a water-cooled furnace wall, expansion of the melting region is suppressed by cooling and then a skull layer is formed, which consists of a non-melted layer and a re-solidified layer or a sintered layer. Since the melted part does not make direct contact with the furnace wall, melting without contamination is then possible due to the existence of this skull and it is also not necessary to choose furnace materials. The research on the High Frequency Induction Skull Melting method that makes use of the electric conductive properties of melted materials was started in 1960 by Sterling and Warren [33], and was developed to a practical stage in the early 1970s by Aleksandrov et al. [34]. Today, this apparatus has been adopted as a crucible for unidirectional solidification processes and the Czochralski method, which are used for the process producing single crystals of MgO, TiO_2, Fe_3O_4, $Nd_3Ga_5O_{12}$, ZrO_2 and so on [34, 35]. Moreover, as this method has the merit of no contamination, attention has been

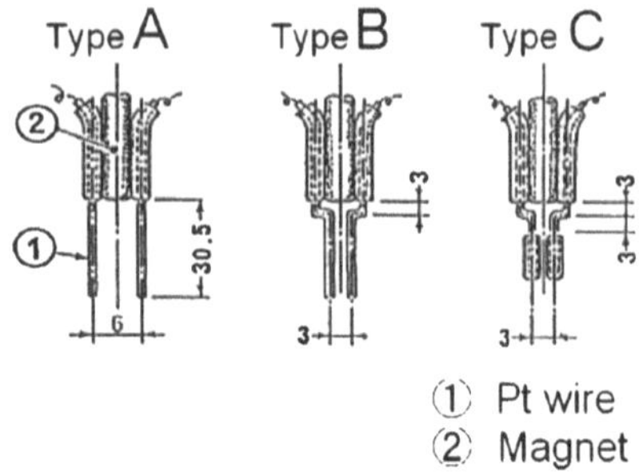

Fig. 4.19 Shape of experimental probes [39]

paid to the melting of high-grade glasses and enclosing of radioactive wastes in glass containers. In this method, a water-cooled coil equipped for magnetic field generation is directly used as a crucible or the cold crucible (see Sect. 2.1.3) is used. In this melting process, a temperature field, a magnetic field, and a velocity field are closely related with each other, so that a reasonable design for this process cannot be obtained without grasping the characteristics of the process parameters. So far, only a few reports [36, 37] exist for the theoretical analysis of this process which is still in an unsatisfactory situation.

2.9 Flow Velocity Detecting Function

When an electrically conductive fluid moves at velocity $\boldsymbol{v}$ under a magnetic field $\boldsymbol{B}$ generated by a permanent magnet, an electromotive force U is caused according to Fleming's right-hand rule

$$U \propto (\boldsymbol{v} \times \boldsymbol{B}) \cdot \boldsymbol{l} \tag{4.23}$$

Here, $\boldsymbol{l}$ is a vector of length with direction. On the basis of this principle, by using a value of U directly measured between dipped electrodes, and a correction coefficient k determined by drawing an inspection line, a velocity v can be obtained from Eq. 4.24,

$$v = k \cdot d \cdot B \cdot U, \tag{4.24}$$

where d is distance between electrodes. This is the principle of the Vivés probe [38]. Hosoya et al. [39] made three kinds of probes as shown in Fig. 4.19 and examined their characteristics. This result is shown in Fig. 4.20. Although these are compact probes with simple structure, local velocity can be obtained to relatively high accuracy. The probe, however, cannot be applied to high temperature processes exceeding the Curie point of permanent magnets. Recently, Gerbeth et al. [40] developed three kinds of contactless flow rate sensors. Those are promising for flow controlling of metals with high melting point such as molten steel.

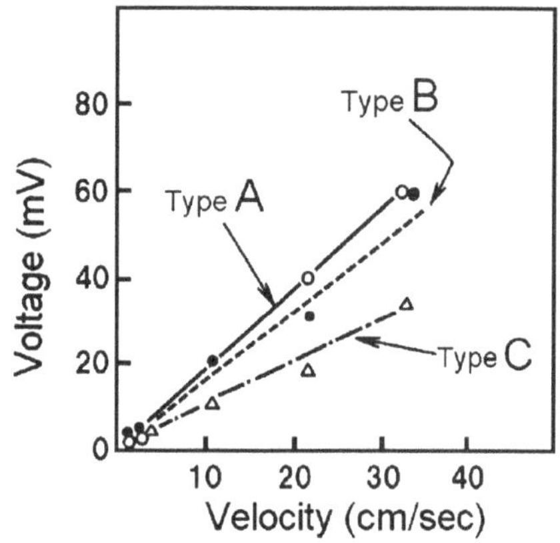

Fig. 4.20 Linear relation between flow velocity and electric voltage [39]

3 Classification of Electromagnetic Processes for Materials

On the basis of imposition methods of electric and magnetic fields, electromagnetic processes of materials can be classified as follows:

- Imposition of magnetic field
  - DC magnetic field
    - Magnetic Czochralski method
    - Electromagnetic brake
    - Shape control of molten metal
    - Shape control of side edge of ribbon produced by twin-roll method
  - AC traveling magnetic field with several Hz ~ 60Hz
    - Electromagnetic stirring in continuous casting
    - ASEA-SKF
    - Electromagnetic pump
    - Rotary type of electromagnetic atomization
    - Control of nozzle flow rate
  - AC magnetic field with 60~ several mega Hz
    - High frequency induction furnace
    - Electromagnetic casting
    - Cold crucible
    - Levitation melting
    - Filming of molten metal

- Imposition of electric current
  - Direct current
    - VAR(Vacuum Arc Remelting)
    - ESR(Electro-Slag Remelting)
    - Arc welding
    - Ohmic heating
    - Aluminum refining oven
    - Plasma
  - Alternating current
    - VAR
    - ESR
    - Arc welding
    - Ohmic heating
    - Plasma

- Simultaneous imposition of magnetic field and electric current
  - Direct current and static magnetic field
    - Electromagnetic stirrer under static magnetic field
    - Horizontal electromagnetic casting
    - Frequency controlling of bubbling
    - Electromagnetic stirrer of ESR
    - Refinement of solidification structure
  - Alternating current and alternating magnetic field — Magnetic mixing in welding
  - Direct current and alternating magnetic field
    - Flow suppression in ESR and VAR
    - Frequency control of bubbling
  - Alternating current and static magnetic field
    - Electromagnetic stirring in ESR
    - Refining of solidified structure

References

1. T. Kozuka, S. Asai, I. Muchi, Tetsu-to-Hagane **75**(3), 470 (1989)
2. M. Garnier, R. Moreau, J. Fluid Mech. **127**, 365 (1893)
3. F. Yamao, K. Sassa, K. Iwai, S. Asai, Tetsu-to-Hagane **83**(1), 30 (1997)
4. S. Amano, K. Iwai, S. Asai, ISIJ Int. **37**(10), 962 (1997)
5. Z.N. Gezelev, U.S. Patent 3,467,166 (1969)
6. T.R. Pritchett, Light Met. Age **10**, 12 (1981)

7. S. Asai, *The 129,130th Nishiyama Memorial Technical Lecture* (The Iron and Steel Institute of Japan, Tokyo, 1989), p. 51 (In Japanese)
8. T. Li, S. Nagaya, K. Sassa, S. Asai, Met. Trans. **26B**(April), 353 (1995)
9. T. Li, K. Sassa, S. Asai, Tetsu-to-Hagane **82**(3), 197 (1996)
10. JRCM NEWS, No.142 (1998), p. 4
11. E. Takeuchi, H. Tanaka, H. Kajioka, *Proceedings of International Symposium on Electromagnetic Processing of Materials* (ISIJ, Nagoya, 1994), p. 364
12. K. Hoshi, T. Suzuki, T. Okubo, N. Isawa, *Extended Abstracts Electrochemical Society, Spring Meeting*, vol. 8-1 (Electrochemical Society, Pennington, 1980), p. 811
13. J.A. Shercliff, J. Fluid Mech. **38**, 353 (1969)
14. I.S. Robinson, J. Fluid Mech. **69**, 475 (1975)
15. S. Chandrasekhar, *Hydrodynamic and Hydromagnetic Stability* (Oxford Press, London, 1961), p. 457
16. T. Kozuka, S. Asai, I. Muchi, Tetsu-to-Hagane **74**(12), 2278 (1988)
17. T. Kozuka, M. Kinoshita, S. Asai, I. Muchi, Tetsu-to-Hagane **76**(10), 1696 (1990)
18. D. Leenov, A. Kolin, J. Chem. Phys. **22**, 683 (1954)
19. E.F. Bepme, *Electromagnetic Casting and Processing of Molten Metal* (Japan Soviet News Agency, 1968), p. 113
20. Ph. Marty, A. Alemany, *Proceeding of Symposium of IUTAM* (The Metal Society, 1984), p. 245
21. P. Park, A. Morihira, K. Sassa, S. Asai, Tetsu-to-Hagane **80**, 389 (1994)
22. A.D. Patel, N. El-Kaddah, *Proceedings of International Symposium on Electromagnetic Processing of Materials* (ISIJ, Nagoya, 1994), p. 115
23. V.M. Korovin, Magnetohydrodynamics **21**, 321 (1985)
24. S. Taniguchi, J.K. Brimacombe, Tetsu-to-Hagane **80**(1), 24 (1994)
25. Y. Tanaka, K. Sassa, K. Iwai, S. Asai, Tetsu-to-Hagane **81**(12), 1120 (1995)
26. F. Yamao, K. Iwai, S. Asai, Tetsu-to-Hagane **83**(1), 30 (1997)
27. C. Vivés, Met. Mater. Trans. B **27**, 445 (1996)
28. C. Vivés, Met. Mater. Trans. B **27**, 457 (1996)
29. J. Saneyoshi, K. Kikuchi, O. Kumamoto, *Handbook of Ultrasonic Technique* (Nikkan Kogyo Shinbun, Ltd., Tokyo, 1978)
30. A.K. Varma, N. El-Kaddah, *Magnetohydrodynamics in Process Metallurgy* (TMS, Warrendale, 1991), p. 299
31. A. Alemany, J. Barbet, Y.R. Fautrelle, R. Moreau, French patent No. 7,717,296 (1977)
32. A.F. Kolesnichenko, I.B. Kazachkob, B.O. Bodyanjuk, N.B. Iysak, *Kapillyarnye MGD Techniya so Svobodnymi Granitsami* (Naukoba Dymka, Kiev, 1988)
33. H.F. Sterling, R.W. Warren, Nature **25**, 745 (1961)
34. V.I. Aleksandrov, V.V. Osiko, A.M. Prokhorov, V.M. Takarintsev, Vestn. Akad. Nauk SSSR **12**, 29 (1973)
35. J.F. Wenckus, J. Cryst. Growth **128**, 13 (1993)
36. B. Caillault, Y. Fautrelle, R. Perrier, J.J. Aubert, in *Liquid Metal Magneto-hydrodynamics*, ed. by J. Lielpeteris, R. Moreau (Kluwer, Dordrecht, 1989), p. 241
37. N. Takasu, K. Sassa, S. Asai, Tetsu-to-Hagane **77**, 496 (1991)
38. R. Ricou, C. Vivés, Int. Heat Mass Transf. **25**(10), 1579 (1982)
39. K. Hosoya, H. Nakato, K. Saito, M. Oguchi, H. Okuda, T. Kayano, Document of Basic Research Committee of Electromagnetic Metallurgy, No.5-5, Feb 1987
40. D. Buchenau, G. Gerbeth, J. Priede, in *Proceedings of 6th International Conference on Electromagnetic Processing of Materials*, Dresden, 2009, p. 383

Chapter 5
Materials Processing by Use of a High Intensity Magnetic Field

Non-magnetic substances such as water, plastic, wood, etc. can be levitated when a magnetic field over 20 T is imposed [1]. This phenomenon is based on the established fact that a magnetization force, which is a well-known force attracting an iron to a magnet, is significantly intensified by a high magnetic field. Thus, in recent times, much attention has been paid to this force. As an example, a super-conducting magnet with a cryostat, which does not require liquid helium as a coolant, has been developed so that a highly intensified magnetic field, as much as 10 T, has become easily available in ordinary laboratories in universities. Effects of a high magnetic field have been examined in a large number of natural science fields such as physics, chemistry and biology and have found many new and interesting phenomena that can not be observed under the ordinary intensity of a magnetic field provided by electric or permanent magnets. For example, Fig. 5.1 shows a water surface depressed by imposition of a high magnetic field. This phenomenon is called the Moses effect for the escape from Egypt story written in the Old Testament [2]. As the second example, Fig. 5.2 shows a living frog being levitated in the bore of a super-conducting magnet, where a gravity force is balanced with a magnetization force [3]. Furthermore, Fig. 5.3 shows that the flame of a candle is deformed by a magnetic field with a gradient. In fact, this phenomenon was first found by Faraday in the nineteenth century and has been understood as an effect of the magnetization force. In addition, various interesting phenomena, such as that the vaporizing rate of water is accelerated and the absorption rate of oxygen gas into water is increased by the imposition of a high magnetic field, have been reported [4, 5]. These circumstances have in recent years given further development of the concept of "Magneto-Science", a subject of research that impacts a variety of science in which high magnetic fields are significant. Reported phenomena relating to "Materials Science" have provided useful information on the creation of new materials, leading finally to the combined identification of "Electromagnetic Processing of Materials" [6]. Of course the Materials Science relating to a high magnetic field is obviously based on a number of principles of physics such as the magnetization force, the Lorentz force, the Zeeman effect, etc., and these principles are combined in complex ways in both physical and chemical phenomena.

S. Asai, *Electromagnetic Processing of Materials*, Fluid Mechanics and Its Applications 99, DOI 10.1007/978-94-007-2645-1_5,
© Springer Science+Business Media B.V. 2012

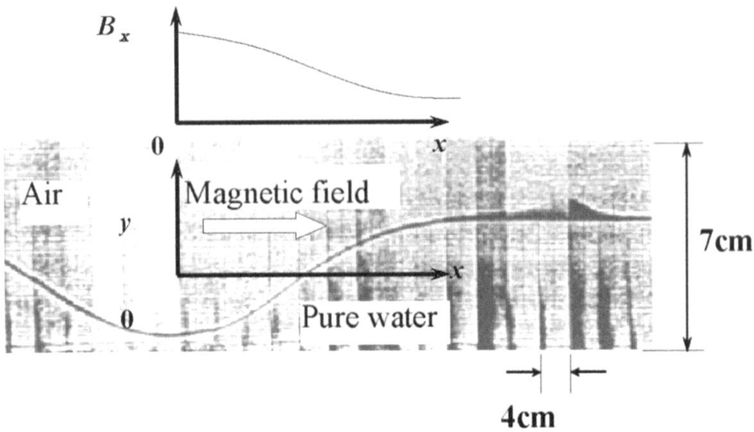

Fig. 5.1 Moses effect (a water shape under a magnetic field [2])

Fig. 5.2 A frog levitated in a bore of a 16 T super conducting magnet [3]

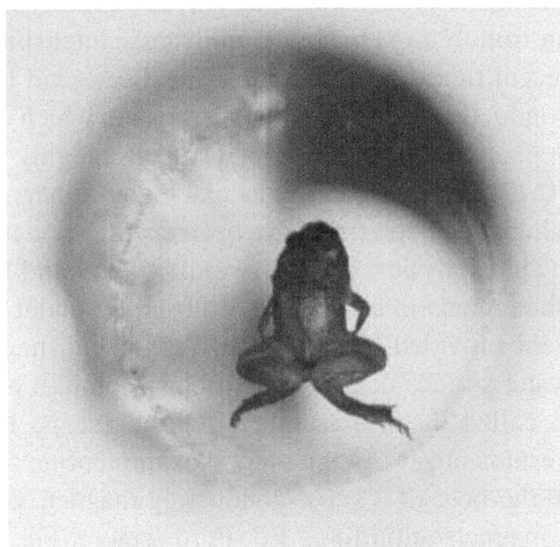

1 Magnetization Force

1.1 Magnetization Force in Non-magnetic Materials

In general the magnetization force is recognized as the force with which a magnet attracts an iron. Applications of the magnetization force given in Eq. 5.1 in industries have been limited to ferromagnetic materials such as an iron.

$$\boldsymbol{F} = \mu_{m0}(\boldsymbol{M} \cdot \nabla)\boldsymbol{H} = (\chi/\mu_{m0})(\boldsymbol{B} \cdot \nabla)\boldsymbol{B} \propto \boldsymbol{B}^2, \tag{5.1}$$

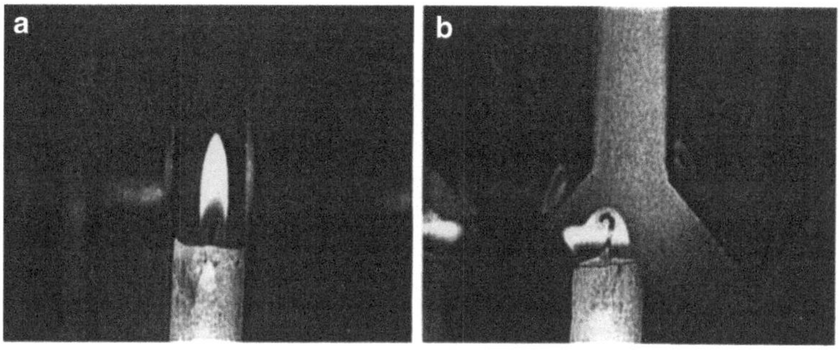

Fig. 5.3 The shape of a candle flame. **a** magnetic field without gradient, **b** magnetic field with gradient

Table 5.1 The relation between magnetization force and magnetic field in ferromagnetic and non-magnetic materials

Magnetization force: $F = (\chi/\mu_{m0})(\boldsymbol{B}\cdot\nabla)\boldsymbol{B} \propto B^2$
Magnetic susceptibility of non-magnetic materials: $\chi_n = 10^{-3}$
Apparent magnetic susceptibility of ferromagnetic materials:
$\chi_m = 10^3$
Thus, $\chi_n/\chi_m = 10^{-6}$
Increase of magnetic flux intensity: 10^3 times(0.001 T → 10 T)
$F \propto B^2 = (10^3)^2 = 10^6$
$\therefore F(\chi_n, 10\text{T}) \approx F(\chi_m, 0.01\text{T})$

where $\boldsymbol{M}$ is the magnetic moment, μ_{m0} the magnetic permeability in a vacuum, χ the magnetic susceptibility. The reason why the application of magnetization force is limited, results in $M_n/M_m = 10^{-6}$, which means the ratio of the value of a magnetic moment in non-magnetic materials M_n to that in ferromagnetic materials M_m is extremely small. If the expression of $\boldsymbol{M} = \chi\boldsymbol{H}$, which holds in only non-magnetic materials could have extended to ferromagnetic materials, we could understand this reason as follows. That is, as $\chi_m \approx 10^3$ for ferromagnetic materials and $\chi_n \approx 10^{-3}$ for non-magnetic materials could be valid, so that about 10^{-6} times difference exists in their ratio. Thus, the magnetization force acting on non-magnetic materials has been neglected, hitherto. By the way, a high magnetic field covering a comparatively large space has been available now without a superconducting magnet using liquid helium and the available intensity of a magnetic field has jumped to 10^3 times (which corresponds to 10 T on the base value of 0.01 T) (see Table 5.1). Thus, the magnetization force in proportion to the square of $\boldsymbol{B}$ is increased to 10^6 times. This fact can compensate for the difference of 10^{-6} in χ (see Table 5.1). That is, roughly speaking, a high magnetic field (10 T) can attract non-magnetic materials similar to the way in which a toy magnet (0.01 T) attracts an iron. The substances listed in the left column of Table 5.2 were levitated when the product of magnetic flux density B and its gradient, BdB/dz that is in proportion to the magnetization force, reached the listed values [7].

Table 5.2 The values of magnetic flux density and the product of magnetic flux density and its gradient for levitating substance [7]

Substance	B(T)	BdB/dz (T^2/m)
Water	27	3,000
Ethanol	21	1,600
Acetone	22	2,000
Bismuth	15.9	730
Antimony	18.8	1,200
Wood	21.5	1,700
Plastic	22.3	2,000

As magnetic susceptibility is different in each crystal orientation, control of crystal alignment should be possible in metals, ceramics and organic materials. Based on this view-point, research has been started into a new materials science direction that makes use of the magnetic field along with temperature, concentration and pressure as common process control parameters [7–14]. Furthermore, it has been known for some time that the magnetic field affects the reaction ratio in radical reactions in organic materials.

1.2 Moses Effect and Inverse Moses Effect

We introduce here the mathematical description of the Moses effect (see Fig. 5.1) proposed by Hirota et al. [15]. Let us assume that a liquid surface is deformed as shown in Fig. 5.4 b under the magnetic field given in Fig. 5.4 a. As no motion takes place in a static system, the sum of a magnetic and a potential energy should be constant as shown in Eq. 5.2.

$$E = -\mu_{m0}\chi H(x)^2/2 + \rho g h(x) = const. \tag{5.2}$$

Equation 5.3 is obtained by equating the values at points A and B in Fig. 5.4 and rearranging the resulting equation:

$$\Delta h = h(x_A) - h(x_B) = (\mu_{m0}\chi/2\rho g)\{H(x_A)^2 - H(x_B)^2\} \tag{5.3}$$

As the constant value of $\mu_{m0}\chi/2\rho g$ multiplying the term showing the difference in the square of the magnetic field is proportional to χ/ρ, the intensity of the Moses effect is determined by a magnetic intensity H, magnetic susceptibility χ and a liquid density ρ. In the case of water which is a diamagnetic substance, its shape is a concavity as seen in Fig. 5.1. It has been calculated by use of Eq. 5.3 that the surface height difference between impositions of 1 and 0 T is about 0.4 mm in pure water. In the case of concentrated copper sulfuric acid, which is a paramagnetic substance, its shape is convex as in Fig. 5.5. The latter case is called an inverse Moses effect.

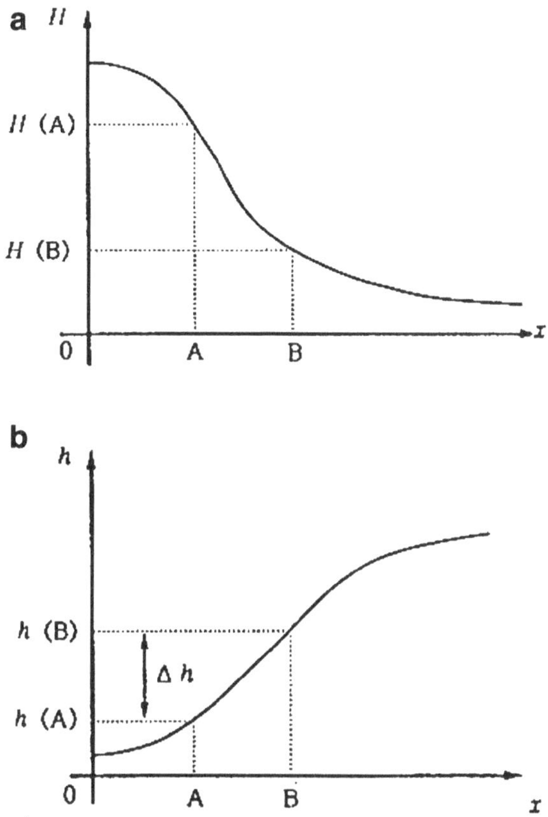

Fig. 5.4 a an assumed magnetic field distribution, b a water surface shape under a magnetic field with gradient [15]

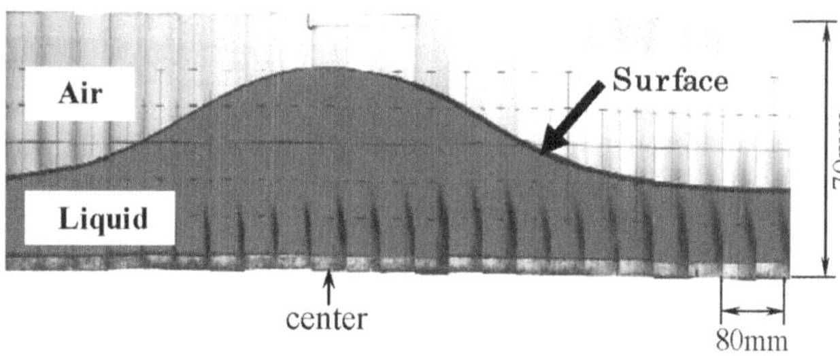

Fig. 5.5 Inverse Moses effect [15]

1.3 Enhanced Moses Effect [15]

Here, the case of two liquid layers is discussed. The total of magnetic and potential energies in each liquid element near the interface between both liquids is derived under the condition given in Fig. 5.6. In order to obtain the shape of the interface, let us imagine the operation that the element of liquid A locating at x_1 is exchanged

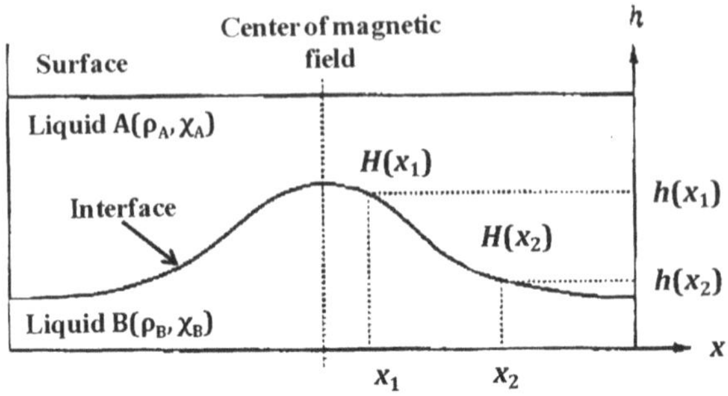

Fig. 5.6 Two laminated liquids located under a magnetic field with gradient [15]

with the element of liquid B at x_2. By repeating this operation the shape of the interface is going to change. The total energy given before and after the exchange is shown as follows:

$$\text{Before exchange}\quad \left[\left\{\rho_A g h(x_1) - \frac{\mu_{m0}\chi_A H(x_1)^2}{2}\right\} + \left\{\rho_B g h(x_2) - \frac{\mu_{m0}\chi_B H(x_2)^2}{2}\right\}\right] dv$$

$$\text{After exchange}\quad \left[\left\{\rho_A g h(x_2) - \frac{\mu_{m0}\chi_A H(x_2)^2}{2}\right\} + \left\{\rho_B g h(x_1) - \frac{\mu_{m0}\chi_B H(x_1)^2}{2}\right\}\right] dv,$$

where ρ_A and ρ_B are the density of liquid A and B, respectively, χ_A and χ_B the magnetic susceptibility of liquids A and B, respectively. Thus, a total energy difference before and after exchanging ΔE is given by Eq. 5.4:

$$\Delta E = \left[(\rho_A - \rho_B)g\{h(x_1) - h(x_2)\} - \frac{1}{2}\mu_{m0}(\chi_A - \chi_B)\{H(x_1)^2 - H(x_2)^2\}\right] dv \tag{5.4}$$

At equilibrium, the difference becomes minimum, that is, $\Delta E = 0$, then Eq. 5.5 is derived from Eq. 5.4.

$$\Delta h \equiv h(x_1) - h(x_2) = \{\mu_{m0}(\chi_A - \chi_B)/2g(\rho_A - \rho_B)\}\{H(x_1)^2 - H(x_2)^2\} \tag{5.5}$$

The coefficient multiplying the term that expresses the difference in the square of the magnetic field is different from that given in Eq. 5.3. That is, the coefficient $(\chi_A - \chi_B)/(\rho_A - \rho_B)$ is the ratio of the differences in magnetic susceptibility and density of both liquids. Thus, it is understood that the interface shape can be enhanced by reducing the density difference even in the case where the magnetic susceptibility difference is small. The picture given in Fig. 5.7 shows the experimental result obtained under the condition where a 0.6 T magnetic field was imposed on transparent monochlorobenzene in the upper layer and a copper sulfuric acid solution in the lower layer. The surface shape of the upper liquid layer is flat,

Fig. 5.7 Enhanced Moses effect [15]

but the interface between the two liquids is deformed as much as 25 mm due to the setting of small density difference within 0.01 g/cm^3 in the two liquids. Hirota et al. [15] gave the name "enhanced Moses effect" to this phenomenon.

2 Magnetism [16]

2.1 Magnetic Classification

Our concept of magnetism is based on our understanding of the motion of electrons and the principle that all substances are composed of atoms that possess electrons. Thus, magnetism essentially exists in all substances. When discussing magnetism, spin angular momentum and orbital angular momentum are essential. This, however, is related to quantum mechanics, so that we do not discuss it in detail here.

There are several ways of classifying magnetism. Here, we briefly classify substances into strong magnetism, weak magnetism and diamagnetism as shown in Table 5.3. All substances except the magnetic materials are called non-magnetic materials. In strong magnetism, we include ferromagnetism, in which all of the magnetic moment in atoms align in the same direction, and ferrimagnetism, in which the magnetic moment in neighboring magnetic atoms are aligned in the opposite direction, but the strong magnetism appears due to the large total absolute values of the magnetic moment. In weak magnetism, we include anti-ferromagnetism, in which the magnetic moment in neighboring atoms compensate each other, and paramagnetism, in which the magnetic moment is induced in the same direction and proportion to an external magnetic field, but its value is very small. Diamagnetism is the classification in which the magnetic moment appear in the opposite direction to the external magnetic field.

2.2 Magnetic Susceptibility

Magnetic susceptibility χ is a characteristic value indicating the degree of ease with which substance can be magnetized. Its definition, however, has not been

Table 5.3 Classification of magnetism

Strong magnetism	Ferromagnetism Ferrimagnetism	Magnetic materials
Weak magnetism	Anti-ferromagnetism Para-magnetism	Non-magnetic materials
Diamagnetism		Non-magnetic materials

universally agreed, due to historical and practical complications, so that several definitions now exist. It is essentially defined as the ratio of magnetization M to magnetic intensity H.

$$\chi \equiv M/H \tag{5.6}$$

Here, various definitions of magnetic susceptibility are classified and the differences among them are shown. The constitutional equation describing the relation between magnetic flux density and magnetic field is shown in Eq. 5.7.

$$B = \mu_m H, \tag{5.7}$$

where μ_m is magnetic permeability.

Magnetic permeability in a vacuum μ_{m0} is given in Eq. 5.8 in the SI(MKSA) unit system and $\mu_{m0} = 1\,(-)$ in the cgs Gauss unit system.

$$\mu_{m0} = 4\pi \times 10^{-7}\,(H/m) \tag{5.8}$$

The dimension of magnetic susceptibility depends on one configuration of the three constitutional equations as given in Table 5.4. Thus, we have to make sure which definition is adopted. In the case of the unit system for E-H, the non-dimensional coefficient χ_r defined as Eq. 5.9 is called a relative susceptibility.

$$\chi_r = \chi'/\mu_{m0}\,(-), \tag{5.9}$$

where χ' is the magnetic susceptibility in the unit system for E-H.

The relations among the magnetic susceptibilities are given in Eq. 5.10.

$$\chi\,(\text{for E} - \text{B}) = \chi_r\,(\text{for E} - \text{H}) = 4\pi\chi''\,(\text{for cgs Gauss}) \tag{5.10}$$

In practice, the cgs Gauss unit is most often used. Furthermore, the magnetic susceptibility per unit mass χ_g (mass magnetic susceptibility) and the magnetic susceptibility per unit molecule χ_m (molar magnetic susceptibility) are used as the individual investigator finds convenient. As an example, the dimensions of χ''_g and χ''_m given in cgs Gauss unit are listed in Table 5.5. In this book, the SI unit system for E-B is adopted so that the relations among the constitutional equations are given as Eq. 5.11.

Table 5.4 The definitions of constitutional equation and magnetic susceptibility [16].

Unit system		Constitutional equation	Definition of magnetic susceptibility	Dimension of magnetic susceptibility
SI	For E-B	$B = \mu_{m0}(H + M)$	$\chi \equiv M/H$	$\left[\frac{A/m}{A/m}\right]$ = non-dimension
	For E-H	$B = \mu_{m0}H + I$	$\chi' \equiv I/H$	$\left[\frac{(wb/m^2)}{A/m} = \frac{(HA/m^2)}{A/m}\right] = \left[\frac{H}{m}\right]$
cgs	cgs Gauss	$B = H + 4\pi I$	$\chi'' \equiv I/H$	$\left[\frac{(emu/cm^3)}{Oe}\right]$ = non-dimension

Table 5.5 Various magnetic susceptibilities where the dimension are given in cgs Gauss units [16]

Name	Nomenclature	Dimension	Common unit
Volume magnetic susceptibility	χ''	[non-dimension]	$emu/(cm^3 \cdot Oe)$
Mass magnetic susceptibility	$\chi''_g = \chi''/\rho$	$\left[\frac{non-dimension}{g/cm^3} = \frac{cm^3}{g}\right]$	$emu/(g \cdot Oe)$
Molar magnetic susceptibility	$\chi''_m = \chi''_g \times m_{mol}$	$\left[\left(\frac{cm^3}{g}\right) \times \left(\frac{g}{mol}\right) = \frac{cm^3}{mol}\right]$	$emu/(mol \cdot Oe)$

ρ density (g/cm^3), m_{mol} molecule mass (g/mol)

$$\boldsymbol{B} = \mu_{m0}(\boldsymbol{H} + \boldsymbol{M}) = \mu_{m0}(1 + \chi)\boldsymbol{H} = \mu_{m0}\mu_r\boldsymbol{H} = \mu_m\boldsymbol{H}, \tag{5.11}$$

where μ_r (= $1 + \chi$) and μ_m (= $\mu_{m0}\mu_r$) are called relative magnetic permeability and magnetic permeability, respectively.

2.3 *Magnetic Anisotropy*

The factors determining the magnetization process are magnetic anisotropy, magnetic-field-induced strain, static magnetic energy and exchange interaction. These factors very much depend on crystal structures (triclinic, monoclinic, orthorhombic, hexagonal, rhombohedra, tetragonal, cubic, etc.), manufacturing processes (melting and solidification, vapor deposition, spattering, rolling, crushing, sintering, etc.), heat treatments (quick cooling, slow cooling, annealing, age hardening, cooling in a magnetic field, etc.) and shapes (bar, thin film, plate, powder, etc.).

Here, magnetic anisotropy is classified into the following four categories: (1) crystal magnetic anisotropy, (2) shape magnetic anisotropy, (3) exchange interaction anisotropy and (4) magnetically induced anisotropy. Those are briefly discussed in the following.

(1) **Crystal magnetic anisotropy:** It expresses the easiness and difficulty of magnetization appearing with respect to crystal structures. These result from the shape anisotropy of an electron cloud in a unit crystal.

(2) **Shape magnetic anisotropy:** It results from the fact that when a magnetic field is imposed on a substance, another magnetic field is induced in the substance in the direction against the imposed magnetic field (demagnetizing field). That is, when the substance is magnetized, the plus and minus poles, which appear at both edges of the substance, induce the demagnetizing field. The demagnetizing field H_d is in proportion to the magnetization of the substance and a coefficient N defined by Eq. 5.12, which is called a demagnetizing factor.

$$H_d = NM \tag{5.12}$$

The factor N is a non-dimensional number and determined by only the shape of the substance. Among the components of N in x-, y- and z-directions, the relation of Eq. 5.13 holds in the SI unit system.

$$N_x + N_y + N_z = 1 \tag{5.13}$$

In the case of a spherical shape, each component of N is isotropic so that $N_x = N_y = N_z = 1/3$. In a plane with infinite length in the x- and y-directions and thin thickness in the z-direction, the distances between the poles of the demagnetizing magnetic field in the x- and y-directions are infinite. The magnetic field induced by magnetic poles decreases in proportion to the square of the distance so that $N_x = N_y = 0$ and $N_z = 1$ are valid.

(3) **Exchange interaction anisotropy:** It appears when ferromagnetic and anti ferromagnetic substances exist next to each other. The origin of a magnetic hysteresis curve shifts from an origin and the coercive force of one substance more strongly appears than that of the other.

(4) **Magnetically induced anisotropy:** It is artificially produced in processes such as precipitation, solidification, heat treatment, rolling, etc. It comes out due to the residual stress and realignment of atoms by rolling, and the residual thermal stress caused by the temperature difference between deposited film and a substrate in a vapor deposition.

3 Magnetization Energy

3.1 Derivation of Magnetization Energy

When a material is subjected to a magnetic field, the material is magnetized and has magnetization energy. Magnetization energy depends on the shape and crystal orientation of the material so that precipitated substances or crystals rotate so as to have lower magnetization energy in the magnetic field. This is a principle of magnetic crystal alignment. Magnetization energy U is defined by Eq. 5.14 [17].

$$U = -\int_0^H \mu_{m0} M dH, \tag{5.14}$$

where the magnetic moment M and the magnetic field induced in a substance H_{eff} are defined by taking account of a demagnetizing field as in Eqs. 5.15 and 5.16.

$$M = \chi H_{eff} \tag{5.15}$$

$$H_{eff} = H_{ex} - NM, \tag{5.16}$$

where N is the demagnetizing factor and H_{ex} the magnetic field imposed from outside of the substance.

(1) **Non-magnetic substance** ($|\chi| \ll 1$, $\chi > 0$ for a paramagnetic substance and $\chi < 0$ for a diamagnetic one)

From Eqs. 5.15 and 5.16, Eq. 5.17 is obtained.

$$H_{eff} = \{1/(1 + N\chi)\} H_{ex} \tag{5.17}$$

Thus, the magnetization energy U is derived from Eq. 5.14 as follows:

$$U = -\int_0^{H_{ex}} \mu_{m0} \frac{\chi H_{ex}}{1 + \chi N} d\left(\frac{H_{ex}}{1 + \chi N}\right) = -\mu_{m0} \chi H_{ex}^2 / \{2(1 + \chi N)^2\} \tag{5.18}$$

(2) **Ferromagnetic substance** ($\chi \gg 1$)

By assuming that the magnetizing curve for a ferromagnetic substance increases linearly with the imposed magnetic field and is saturated to the magnetic moment M_s at H_s as shown in Fig. 5.8, the magnetization energy U is obtained as Eq. 5.19 in the same derivation method given in Eq. 5.18.

$$\begin{aligned} U &= -\int_0^{H_{ex}} \mu_{m0} M d(H_{ex} - NM) \\ &= \int_0^{H_s} \mu_{m0} \{\chi / (1 + \chi N)\} H_{eff} d \{\chi / (1 + \chi N)\} H_{eff} \\ &\quad - \int_{H_s}^{H_{ex}} \mu_{m0} M d(H_{ex} - NM) \\ &= -\{\mu_{m0} \chi H_s^2 / 2(1 + \chi N)^2\} - \mu_{m0} M_s \int_{H_s}^{H_{ex}} dH_{ex} + \mu_{m0} M_s N \int_{M_s}^{M_s} dM \\ &= -\{\mu_{m0} M_s H_s / 2(1 + \chi N)^2\} - \mu_{m0} M_s H_{ex} + \mu_{m0} M_s H_s \end{aligned} \tag{5.19}$$

Fig. 5.8 Magnetizing curve of a ferromagnetic substance

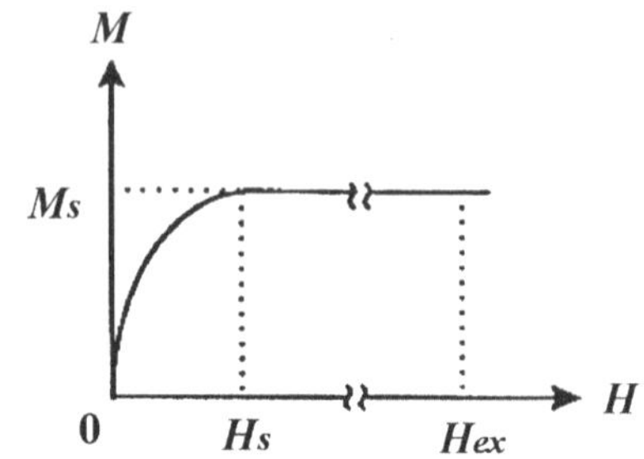

3.2 Shape Magnetic Anisotropy

When a solid phase having a magnetic susceptibility different from a molten phase precipitates during solidification, we have to take into account the molten phase having existed in the place where the solid phase precipitates. The demagnetizing factor N of the solid phase depends on its shape. Thus, not only the magnetic susceptibility of the solid and molten phases, but also its shape should be considered to evaluate the change of the magnetization energy during solidification. In the case where a ferromagnetic substance precipitates in a non-magnetic molten phase, the magnetization energy accompanied with this phase transformation is written in Eqs. 5.20 and 5.21 by the use of Eqs. 5.18 and 5.19. The subscripts of ($\perp$) and ($\parallel$) in Eqs. 5.20 and 5.21 mean the two cases where a magnetic field is applied in perpendicular and parallel to the longer axis of a precipitated phase, respectively.

$$U_{\perp} = -\mu_{m0} M_s H_{ex} + \left\{ 1 - \frac{1}{2(1+\chi_p N_{\perp})^2} \right\} \mu_{m0} M_s H_s + \frac{\mu_{m0} \chi_{med} H_{ex}}{2(1+\chi_{med} N_{\perp})^2} \tag{5.20}$$

$$U_{\parallel} = -\mu_{m0} M_s H_{ex} + \left\{ 1 - \frac{1}{2(1+\chi_p N_{\parallel})^2} \right\} \mu_{m0} M_s H_s + \frac{\mu_{m0} \chi_{med} H_{ex}}{2(1+\chi_{med} N_{\parallel})^2}, \tag{5.21}$$

where χ_p is the magnetic susceptibility of the precipitated substance and χ_{med} that of the medium.

In the case where the shape of the precipitated substance is a plate or rod and the magnetic field is imposed on the longer axis of the precipitated substance, $N_{\perp} \gg N_{\parallel}$ is. When the precipitated substance is composed of ferromagnetic polycrystals, we can consider $M_{\perp} \approx M_{\parallel}$ Furthermore, the absolute value of the second term in Eqs. 5.20 and 5.21 is considered to be much larger than that of the third term. Thus, the difference between the magnetization energies is

$$\Delta U = U_{\perp} - U_{\parallel} = \left\{ -\frac{1}{2(1+\chi_p N_{\perp})^2} + \frac{1}{2(1+\chi_p N_{\parallel})^2} \right\} \mu_{m0} M_s H_s \tag{5.22}$$

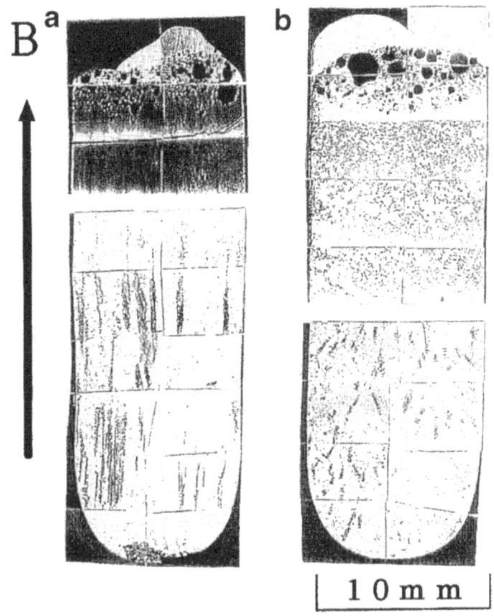

Fig. 5.9 Micro-structure Bi-4mass%Mn alloy. **a** with magnetic field (4.5 T), **b** without magnetic field

By taking account the condition of $N_{\perp} > N_{\parallel}$, we get Eq. 5.23.

$$U_{\perp} > U_{\parallel} \tag{5.23}$$

When a precipitated particle is composed of poly-crystals, the particle should be isotropic. Thus, the particle of a ferromagnetic substance aligns its longitudinal direction to be parallel to the direction of an applied magnetic field.

3.3 Metals Solidification

Regarding crystal orientation in metal solidification, Mikelson et al. [18] first reported that the macrostructure of Al-Cu and Cd-Zn alloys, which are non-magnetic materials, aligned to the direction of the magnetic field during solidification. However, the details on the method for evaluating the orientations of crystals and textures were not clearly written in the report.

Figure 5.9 shows the result of conventional solidification in a non-magnetic molten alloy of Bi-4mass%Mn, where a ferromagnetic inter-metallic compound of MnBi precipitates during solidification. The precipitated dendrites of MnBi distribute at random in the case without magnetic imposition, but in the case with magnetic imposition the primary axes of dendrites align parallel to the direction of the magnetic field. This result implies that the theoretical derivation of Eq. 5.23 agrees with the result in Fig. 5.9.

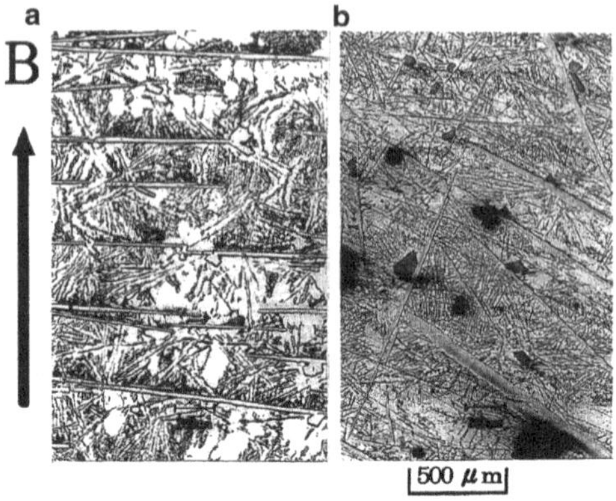

Fig. 5.10 The micro structures of Al-11mass%Si-2mass%Fe alloy. **a** with magnetic field (5 T), **b** without magnetic field

In the case where both the precipitated phase and medium of the matrix are non-magnetic, the magnetization energy evaluated by Eq. 5.18 is given as follows:

$$U_\perp = -\{\mu_{m0}\chi_p H_{ex}^2/2(1+\chi_p N_\perp)^2\} + \{\mu_{m0}\chi_{med} H_{ex}^2/2(1+\chi_{med} N_\perp)^2\} \tag{5.24}$$

$$U_\parallel = -\{\mu_{m0}\chi_p H_{ex}^2/2\,(1+\chi_p N_\parallel)^2\} + \{\mu_{m0}\chi_{med} H_{ex}^2/2(1+\chi_{med} N_\parallel)^2\} \tag{5.25}$$

As the precipitated substance may be poly-crystals, $M_\perp \approx M_\parallel$ is valid, and $|\chi_p| \ll 1$ and $|\chi_{med}| \ll 1$ are due to a non-magnetic substance. Then, the following equation is obtained from Eqs. 5.24 and 5.25:

$$\Delta U = U_\perp - U_\parallel \propto (N_\perp - N_\parallel)(\chi_p^2 - \chi_{med}^2)(\mu_{m0} H_{ex}^2/2) \tag{5.26}$$

When $N_\perp > U_\parallel$, $\Delta U > 0$ holds if $|\chi_p| > |\chi_{med}|$ is valid so that the axial direction of the precipitated phase is parallel to the direction of an imposed magnetic field. On the other hand, $\Delta U < 0$ holds if $|\chi_p| < |\chi_{med}|$ is valid so that the axial direction of the precipitated phase is perpendicular to the direction of a magnetic field. Figure 5.10 shows the experimental result where a non-magnetic inter-metallic compound of Al-9%Si-15%Fe precipitated with a plane shape from a non-magnetic molten metal of Al-11%Si-2%Fe. It is understood from Eq. 5.26 that $|\chi_p| < |\chi_{med}|$ is valid in this system.

Yasuda et al. [19] reported that the crystal orientation of BiMn alloy, which is ferromagnetic material, aligned in the direction of the magnetic field, where they reheated the specimen which was prepared by rapid quenching, up to a liquid and solid zone in a magnetic field.

A chemically active material with a high melting point, $Dy_2Fe_{14}B$, was melted and solidified under the case with and without a magnetic field in a cold crucible equipped in a super conducting magnet. The magnetizing curves obtained under the imposition of a magnetic field indicate a strong magnetic anisotropy [20, 21], as shown in Fig. 5.11.

Problem 5.1 Derive Eq. 5.26.

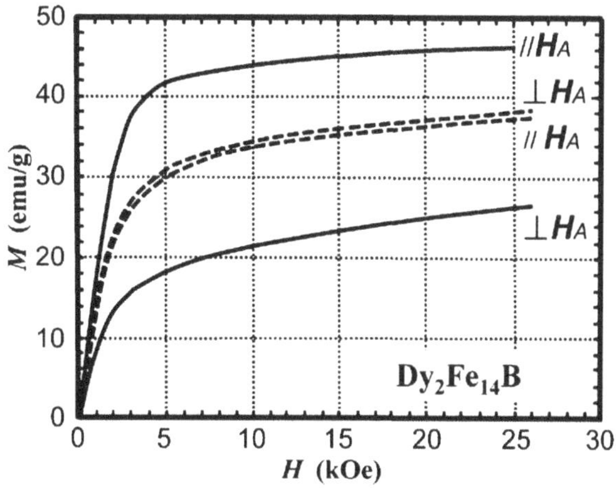

Fig. 5.11 Magnetizing curves of $Dy_2Fe_{14}B$ [21], *solid line*: with magnetic field, *dotted line*: without magnetic field, //: parallel to magnetic field, ⊥: perpendicular to magnetic field

4 Magnetic Crystal Alignment

The imposition of a high magnetic field can control crystal alignment in materials. This principle can be applied to any material composed with anisotropic crystal structures [22–35]. That is, when materials are put in a magnetic field, the crystals exhibiting anisotropic magnetic susceptibility align themselves to the direction of the maximum susceptibility parallel to the magnetic field.

Actually, Asai et al. reported that crystals of metals with magnetic anisotropy such as zinc and bismuth can be aligned in various processes such as solidification [22], electro-deposition [23] and vapor-deposition [24].

4.1 Basic Principle of Magnetic Crystal Alignment

When a substance is put in a magnetic field $\boldsymbol{H}$, a magnetization $\boldsymbol{M}$ will be induced as expressed by Eq. 5.6, which is rewritten as the following equation:

$$\boldsymbol{M} = \chi \boldsymbol{H}, \tag{5.6}$$

where χ is the magnetic susceptibility tensor. For a single crystal with cubic structure, the magnetic susceptibility in each direction of the crystal is equivalent, so that the vector of magnetization $\boldsymbol{M}$ is always parallel to the vector of magnetic field $\boldsymbol{H}$. That is, a crystal with cubic structure can not be aligned by a magnetic field. On the other hand, the magnetic susceptibilities of crystals with crystal anisotropy or shape anisotropy depend on their crystal orientation so that the magnetic susceptibilities in each direction can be denoted as tensors. Let us take an arbitrary rectangular coordinate (χ_1, χ_2, χ_3), denote the components of $\boldsymbol{M}$ as (M_1, M_2, M_3), and write the components of $\boldsymbol{H}$ as (H_1, H_2, H_3), then we can get:

$$\begin{pmatrix} M_1 \\ M_2 \\ M_3 \end{pmatrix} = \begin{pmatrix} \chi_{11} & \chi_{12} & \chi_{13} \\ \chi_{21} & \chi_{22} & \chi_{23} \\ \chi_{31} & \chi_{32} & \chi_{33} \end{pmatrix} \begin{pmatrix} H_1 \\ H_2 \\ H_3 \end{pmatrix}, \tag{5.27}$$

where $\boldsymbol{\chi}$ is a second-order symmetry tensor. Taking the primary axes (χ_1, χ_2, χ_3) into consideration, the above equation can be modified with three diagonal elements as Eq. 5.28.

$$\begin{pmatrix} M_1 \\ M_2 \\ M_3 \end{pmatrix} = \begin{pmatrix} \chi_1 & 0 & 0 \\ 0 & \chi_2 & 0 \\ 0 & 0 & \chi_3 \end{pmatrix} \begin{pmatrix} H_1 \\ H_2 \\ H_3 \end{pmatrix} \tag{5.28}$$

When the crystal structure is a complicated one, the crystal axes may not be in accordance with the primary axes of the tensor. It is, however, consistent for crystals with symmetry structure such as isotropic (cubic), one-axial (tetragonal, trigonal, hexagonal) and two-axial (orthorhombic, monoclinic) crystals.

Let us take a hexagonal crystal as an example; its c-axis is defined as the z-axis, and its a-axis (perpendicular to the c-axis) is defined as the x- or y-axis. The magnetic susceptibility tensor then becomes

$$\begin{pmatrix} M_1 \\ M_2 \\ M_3 \end{pmatrix} = \begin{pmatrix} \chi_a & 0 & 0 \\ 0 & \chi_a & 0 \\ 0 & 0 & \chi_c \end{pmatrix} \begin{pmatrix} H_1 \\ H_2 \\ H_3 \end{pmatrix} \tag{5.29}$$

In most cases, $\chi_a \neq \chi_c$ holds, and their values differ from one substance to another. Thus, when a magnetic field H is imposed, the magnetization energy U in a volume V is

$$U = -(1/2)\mu_{m0}\chi_{\perp}VH^2 - (1/2)\mu_{m0}\Delta\chi VH^2 cos^2\theta \tag{5.30}$$

Here, μ_{m0} is the permeability in vacuum, θ the angle between the easy magnetization axis and the imposed magnetic field, V the crystal volume, $\Delta\chi$ the difference between magnetic susceptibilities in crystal axes, $\chi_{\perp}$ the magnetic susceptibility in the direction perpendicular to the applied magnetic field. The value of the magnetization energy given by Eq. 5.31 which is derived from Eq. 5.18 under the condition of $|\chi| \ll 1$, determines a preferred crystal direction depending on the magnetic susceptibility of each crystal axis.

$$U = -(1/2\mu_{m0})\,\chi_{a,b}\,VB^2 - (1/2\mu_{m0})\,(\chi_c - \chi_{a,b})VB^2\,cos^2\theta, \tag{5.31}$$

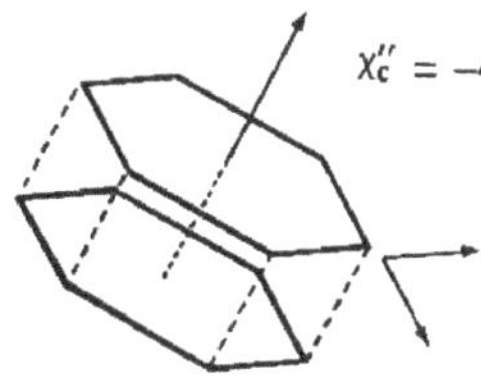

Fig. 5.12 Magnetic susceptibility in each direction of graphite crystal

where χ_c and $\chi_{a.b}$ are the magnetic susceptibilities in the c-axis and the a- or b-axis, respectively. θ is the angle between the directions of a magnetic field and the c-axis of a crystal. When $\chi_c > \chi_{a.b}$, i.e., $U_c = -(\chi_c/2\mu_{m0})VB^2 < U_{a.b} = -(\chi_{a.b}/2\mu_{m0})VB^2$, the c-axis of crystals is the preferred axis parallel to the direction of the imposed magnetic field. In contrast, when $\chi_c < \chi_{a,b}$, i.e. $U_c > U_{a,b}$, the a- or b-axis of the crystal is the preferred axis which is parallel to the magnetic field. That is, the c-axis of the crystal aligns in a plane perpendicular to the imposed magnetic field.

4.2 *Magnetic Alignment of Graphite Crystals*

Graphite, which is a phase of carbon and has a hexagonal structure as shown in Fig. 5.12, has different magnetic susceptibilities depending on its crystal orientation. The normal direction of the c-plane as a basic plane is the c-axis and the normal directions to the c-axis are the a- and b-axes. The values of magnetic susceptibility [36] at room temperature are $\chi''_{\mathrm{a,b}} = -6.7 \times 10^{-7}\,(-)$ and $\chi''_{\mathrm{c}} = -47 \times 10^{-6}\,(-)$. That is, those are about 70 times different among each direction. When a graphite crystal is so set that the c-axis is parallel to a magnetic field direction, the graphite crystal is expected to rotate as shown in Fig. 5.13, since the repulsion force on the c-axis is larger than that on the a- or b-axis. In order to confirm this estimation, graphite powder with a mean particle diameter of 70 μm was added to a liquid resin and mixed. Then, the liquid resin was solidified under the imposition of a 5 T magnetic flux density. It can be estimated that more of the spectra showing the a- and b-planes appear in the cross-section of the resulting specimen cut in perpendicular to the magnetic field direction, and more of those of the c-plane in the cross-section of that cut parallel to it. Figure 5.14 shows the x-ray diffraction (XRD) taken on the cross-sections cut in parallel and perpendicular to the magnetic field direction. On the plane perpendicular to the magnetic field direction, the peaks at $2\,\theta = 43°$ and 77° increase and those peaks decrease at those angles on the parallel plane. The peaks at $2\,\theta = 43°$ and 77° correspond to the (100) and (110) planes which indicate the a- and b-planes. Under the assumption that graphite particles were spherical, the demagnetizing factors in each direction were equal i.e., $N_a = N_b = N_c = 1/3$, the magnetization energy would depend only on the magnetic susceptibility. As graphite is a diamagnetic substance, Eq. 5.18 is reduced to Eqs. 5.32 and 5.33.

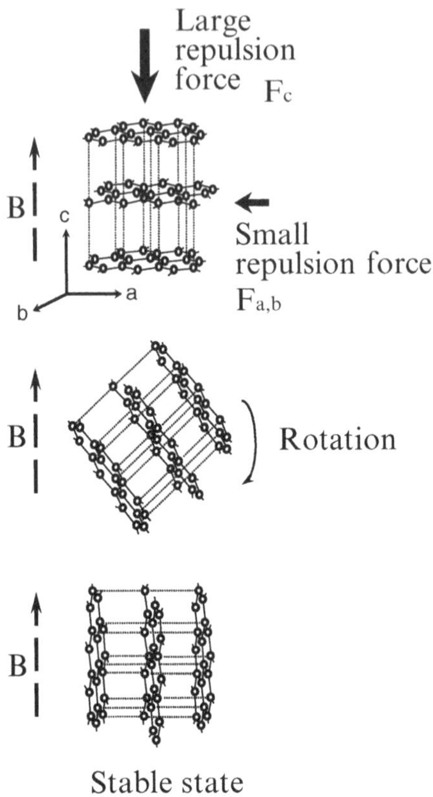

Fig. 5.13 The rotation of a graphite crystal in a magnetic field

$$U_{a,b} = -\mu_{m0}\chi_{a,b}H_{ex}^2/\{2(1+\chi_{a,b}N)^2\} \tag{5.32}$$

$$U_c = -\mu_{m0}\chi_c H_{ex}^2/\{2(1+\chi_c N)^2\} \tag{5.33}$$

As $\chi_c < \chi_{a,b}$, $|\chi_{a,b}| \ll 1$ and $|\chi_c| \ll 1$ are valid, Eq. 5.34 is obtained.

$$U_c > U_{a,b} \tag{5.34}$$

That is, the magnetization energy is smaller in the case where a magnetic field is applied in the a- or b-direction than in the case where it is applied in the c-direction. This indicates the former case is more stable and this theoretical estimation agrees with the experimental results given in Fig. 5.14. Figure 5.15 shows the experimental results obtained by changing the diameter of particles. The smaller the particles are, the more the degree of crystal alignment is intensified. This is understood as follows: as the particles used here are poly-crystals, the number of crystals in the particle decreases with decreasing particle size so that the crystal magnetic anisotropy in the particle are intensified with decreasing particle size.

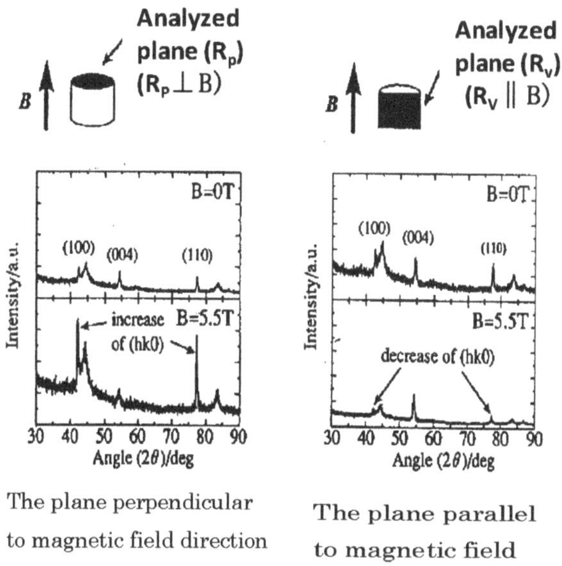

Fig. 5.14 The XRD patterns of graphite particles with and without a magnetic field

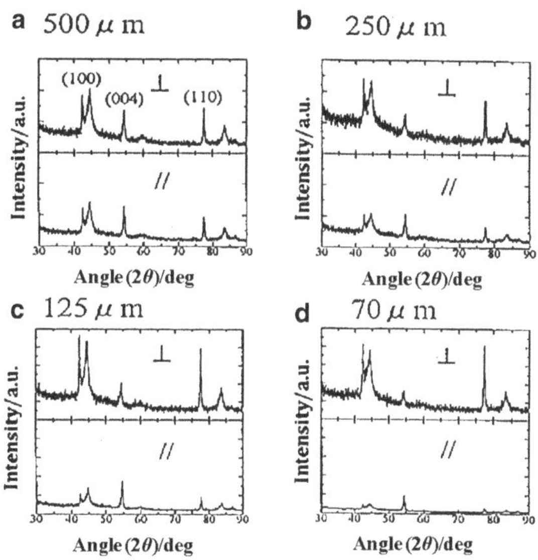

Fig. 5.15 XRD patterns with different size of graphite particles. The *upper figure* shows the plane perpendicular to magnetic field direction (⊥). The *lower figure* the plane parallel to magnetic field direction (//)

4.3 Size Classification [37]

When a slip casting under a magnetic field is used, a colloidal process plays an important role. Thus, let us start the classification of a particle size to understand what external force affects the motion of the particle in a liquid. The crystal alignment is greatly affected by Brownian motion force and gravity force, and which force is dominant is largely dependent on particle size. For a large crystal, the gravity force plays an important role, while the Brownian motion force can be ignored. When a particle is small, the situation is vice versa. When a high magnetic field is introduced, the particle size range which works effectively for the crystal alignment can further be classified as follows:

(1)

$$r_{min} \equiv \left(\frac{3kT\,\mu_{m0}}{2\pi\Delta\chi B^2}\right)^{1/3} < r < r_{cri} \equiv \left(\frac{243kT\mu^2}{8\pi e g^2\rho_p(\rho_p-\rho_l)^2}\right)^{1/7},$$

where k is the Boltzmann constant and μ the viscosity coefficient in a liquid. In this case the Brownian motion is more active than the gravity motion. That is, during alignment of crystalline particles sized in this range, we do not need to consider the gravity force but only the magnetization force.

(2)

$$r_{cri} < r < r_{max} \equiv \sqrt{\frac{3\Delta\chi B^2 L}{4\mu_{m0}g(\rho_p-\rho_l)}},$$

where L is the falling length of particle. In this case the gravity force is more active than the Brownian motion force. In other words, both the gravity force and magnetization force should be taken into consideration.

4.4 Magnetic Slip Casting

A novel process, in which a high magnetic static field is imposed during slip casting was first proposed by Suzuki et al. [38, 39] to fabricate highly crystal aligned ceramics. For magnetic crystal alignment, the following three conditions are required [40]: (1) crystals have an anisotropic magnetic susceptibility, (2) particles exist in a weakly constrained medium, and (3) the crystal size is in a suitable range as mentioned in Sect. 4.3. The functions provided by each different magnetic field are given in the following.

4.4.1 Stationary Magnetic Field

Let us express the magnetic fields which are imposed in parallel and perpendicular to the direction of slip casting, as expressed in Pa and Pe magnetic fields, respectively. The surfaces of top, side 1 and side 2 in a specimen are defined as indicated in Fig. 5.16. When the Pa magnetic field is imposed, side 1 and side 2 become the same so that they refer to the side surface. When the gravity force can be neglected,

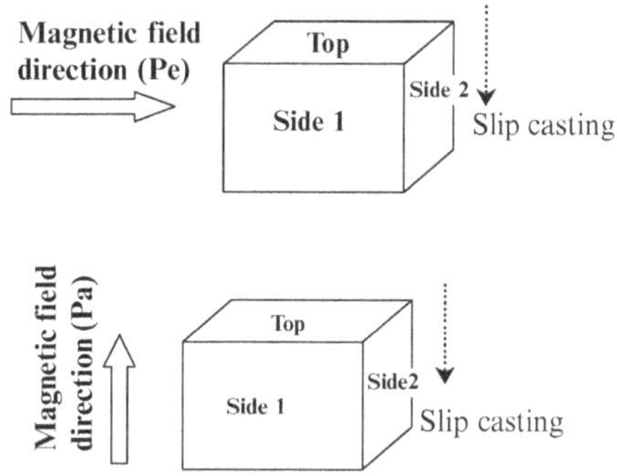

Fig. 5.16 The schematic view of the magnetic field direction and the definition of surface [41]

Table 5.6 The expected crystal orientation of materials with different anisotropic magnetic susceptibility in a slip casting under Pe or Pa magnetic field when the effect of gravity force is not considered

	$\chi_c > \chi_{a,b}$			$\chi_c < \chi_{a,b}$		
Magnetic field direction	Top	Side1	Side2	Top	Side1	Side2
Pe	a,b	a,b	c	a,b,c	a,b,c	(a,b)
Pa	c	a,b	a,b	(a,b)	a,b,c	a,b,c

(a,b) Random distribution of a- and b-planes

the expected crystal alignment of materials under the Pe and Pa magnetic fields is given in Table 5.6. In this table, a, b and c indicate the crystalline planes of materials. It can be seen that if $\chi_c > \chi_{a,b}$, it is relatively easy to get a high crystal alignment because of no freedom in the c-axis direction. If the Pe magnetic field is imposed, the top and side 1 surfaces show the a- or b-plane and the side 2 surface shows the c-plane. If the Pa magnetic field is imposed, the top surface shows the c-plane and the others show the a- or b-plane. In contrast, if $\chi_c < \chi_{a,b}$, it is difficult to get a high crystal alignment because of the freedom in the c-axis direction. If the Pe magnetic field is imposed, the a- or b-plane and the c-plane appear in the top and side 1 surfaces, although the a- or b-plane which distributes randomly appears on the side 2 surface. If the Pa magnetic field is imposed, the a- or b-plane and the c-plane appear on both side surfaces and the a- or b-plane which is randomly distributed appears on the top surface.

4.4.2 Rotating Magnetic Field [41, 42]

When the static magnetic field is imposed on crystals with $\chi_c < \chi_{a,b}$, the c-axis of crystals aligns to arbitrary directions. The alignment of the c-axis of crystals is obtained only when the magnetic field rotates and the direction of the c-axis is perpendicular to the rotating magnetic field plane. Figure 5.17 schematically shows the function of the rotating magnetic field. A magnetic field with a high intensity, which is often needed for the crystal alignment of weak magnetic materials, can

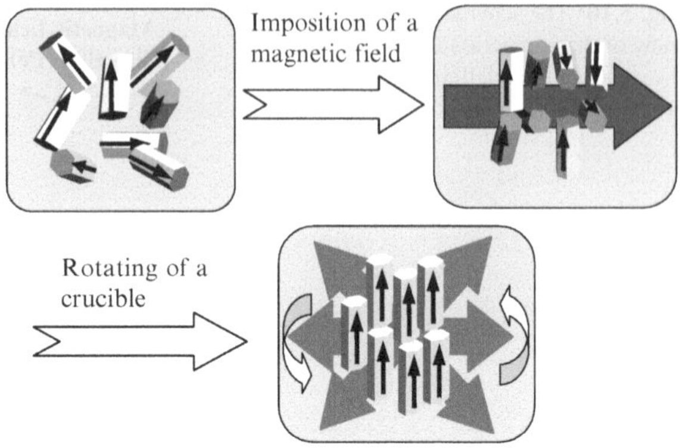

Fig. 5.17 Schematic view of experimental apparatus for rotation of a crucible under a magnetic field

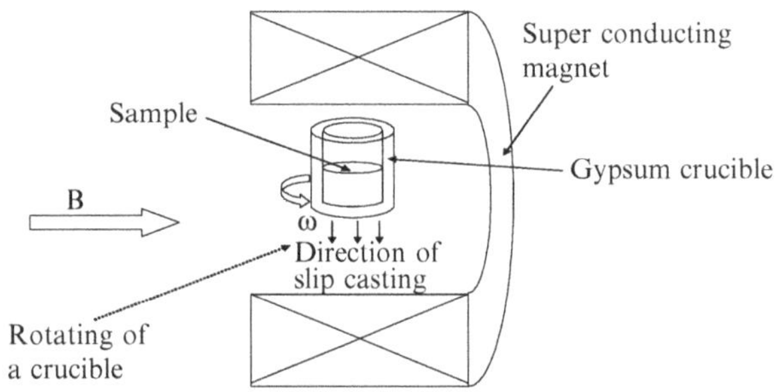

Fig. 5.18 A schematic view showing the experimental setup for rotating a crucible in a stationary magnetic field [41]

Table 5.7 The expected crystal orientation of materials with different anisotropic magnetic susceptibility in a slip casting under Pe or Pa magnetic field

	$\chi_c > \chi_{a,b}$			$\chi_c < \chi_{a,b}$ (with rotation of a crucible)		
Magnetic field direction	Top	Side 1	Side 2	Top	Side 1	Side2
Pe	a,b	a,b	c	c	a,b	a,b
Pa	c	a,b	a,b	a,b	a,b,c	a,b,c

be generated only by a super-conducting magnet. Such a magnetic field, however, cannot be rotated in practice. From the view point of the relative motion, however, the case where a crucible containing crystals is fixed and the magnetic field is rotating is equivalent to the case where the crucible is rotated under a static magnetic field. Thus, the c-axis of crystals was aligned to the direction parallel to the gravity force direction by using such an experimental setup as shown in Fig. 5.18. Under this condition, Table 5.6 can be modified to Table 5.7. This theoretical prediction was confirmed by the experiment, where a crucible was rotated in the static magnetic field during a slip-casting of Si_3N_4 powder, which has a hexagonal structure.

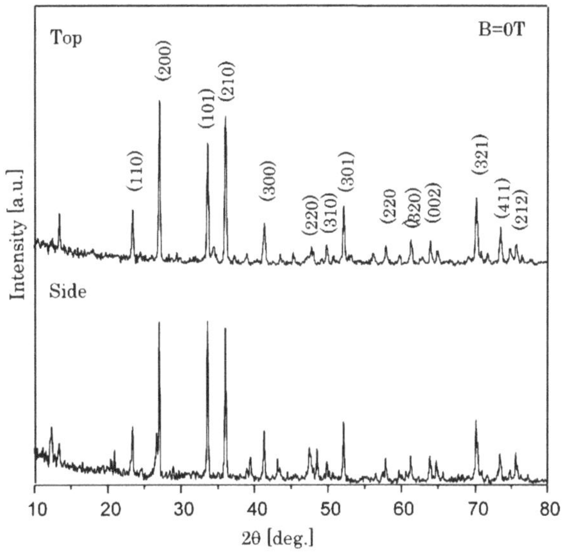

Fig. 5.19 X-ray diffraction patterns of the sample obtained from α-Si_3N_4 powder with β-Si_3N_4 seeds under no magnetic field [41]

4.4.3 Fabrication of Textured Silicon Nitride Ceramics [41]

Figure 5.19 illustrates the XRD profile and the orientation index of the sample that was prepared by the α-Si_3N_4 powder with β- Si_3N_4 seeds, followed by sintering at 1,800C for 1.5 h without a magnetic field. Although the difference between the top and side surfaces of the specimen exits, it is small. On the other hand, when the specimen was rotated under the magnetic field, the peaks clearly changed as seen in Fig. 5.20. In the top surface, the (h k 0) peaks are lower and higher in the side surface. The reason can be explained as follows: when the crucible was rotated, the c-axis of β-Si_3N_4 seeds with rod-like morphology more easily aligned to the direction perpendicular to the magnetic field than the c-axis of α-Si_3N_4 powder did. Moreover, although the rod axis (c-axis) was parallel to gravity direction, the rod could keep the posture because the magnetic force was larger than the gravity force. During the sintering process, the preexisting β-Si_3N_4 seeds highly orientated by the rotating magnetic field provide the preferential nucleation sites of a newly formed β phase in the whole specimen so that the crystal alignment proceeds. Figure 5.21 shows the SEM micrograph of the polished surfaces of the specimen. It also can be seen that β-Si_3N_4 rod grains appear randomly distributed in the specimen which was prepared without exposure to a magnetic field (Fig. 5.21 **a**, **b**). In the case of the specimens prepared with a rotation under the magnetic field, the highly textured material can be seen as shown in Fig. 5.21 **c**, **d**. In Fig. 5.21 **c** where the top surface of the specimen is shown, the hexagonal morphology of the β-Si_3N_4 is observed. On the other hand, the hexagonal rods are exhibited on the side surface as shown in

Fig. 5.20 X-ray diffraction patterns of the sample obtained from α-Si_3N_4 powder with β-Si_3N_4 seeds under rotating magnetic field [41]

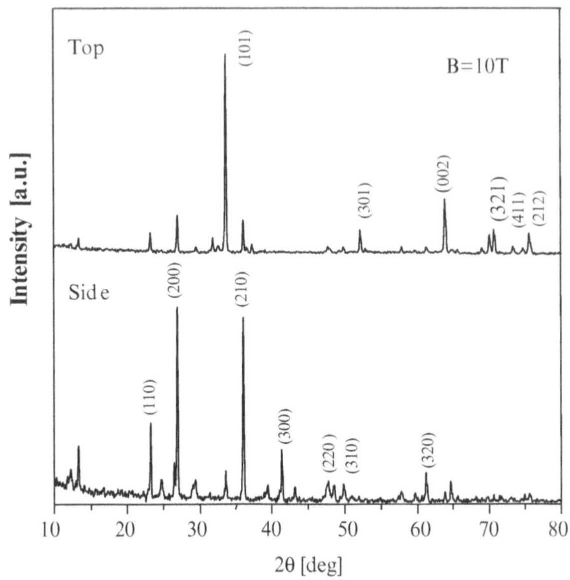

Fig. 5.21 SEM micrographs of specimens made of α-Si_3N_4 powder with β- Si_3N_4 seeds [41]

Fig. 5.21d. The fact that the rod-like β-Si_3N_4 aligns in one direction agrees with the theoretical consideration (the case of $\chi_c < \chi_{a,b}$, Pe magnetic field with rotation of a crucible given in Table 5.7).

4.4.4 Dynamic Behavior of a Crystal in Magnetic Field [43]

When a crystal is free to move, it rotates with its easy magnetization axis parallel to the magnetic field direction so as to decrease the magnetization energy by a magnetization torque, which is derived from Eq. 5.30 as expressed by Eq. 5.35.

$$T = -\partial U/\partial\theta = -(1/2)\,\mu_{m0} V \Delta\chi H^2 \sin 2\,\theta \tag{5.35}$$

When a magnetic field of H is imposed on a spherical single crystal with density ρ and radius r, which is suspended in a solution with viscosity μ and the same density of the crystal, the rotation behavior of the crystal can be expressed by the following differential equation:

$$(2/5)\pi\rho r^5\,(d^2\theta/dt^2) + 8\pi\mu r^3\,(d\theta/dt) = T \tag{5.36}$$

For a non-magnetic crystal in sub-micron order, the inertial force (the first term of the left-hand side in Eq. 5.36) only dominates for a very short period of time so that the corresponding term can be neglected. Then we can get the solution as:

$$\tan\theta = \tan\theta_0\;\exp\,(-t/\tau_s) \tag{5.37}$$

Here, θ_0 is the angle between the easy magnetization axis and the imposed magnetic field at $t = 0$. In Eq. 5.37, τ_s is the time in which the value of tan θ decays from the initial value to its 1/e as given in Eq. 5.38.

$$\tau_s = 6\mu/(\Delta\chi\mu_{m0}H^2) \tag{5.38}$$

It is called the crystal alignment time under a static magnetic field. As shown in Fig. 5.22, let us consider a system where hexagonal crystals with magnetic anisotropy $\chi_c < \chi_a$ are suspended in a solution with viscosity μ. A magnetic field H (t) is imposed in the x-y plane with a rotation speed of ω. Let us define the angle φ (t) between the unit vector c(t) of the c-axis and the z-axis and the angle θ (t) between the orthogonal projection vector of the c-axis in the x-y plane and the x-axis. Then, the rotation motion of the crystal can be described by the following two equations:

$$d\theta/dt = -(1/2\tau_s)\,\cos^2(\varphi - \omega t)\,\sin 2\theta \tag{5.39}$$

$$d\varphi/dt = (1/2\tau_s)\,\sin 2\,(\varphi - \omega t) \tag{5.40}$$

Based on Eqs. 5.39 and 5.40, the rotation behavior of crystals under a time-varying magnetic field can be classified according to the alignment time τ_s.

(1) $\tau_S \rightarrow \infty$ $\;(H(t) \rightarrow 0$ or $\mu \rightarrow \infty)$

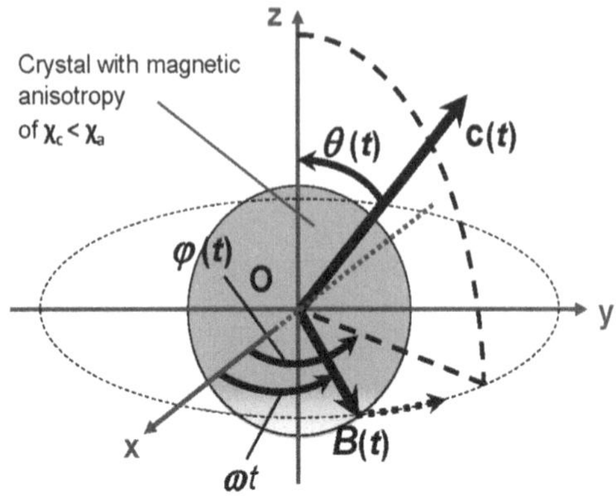

Fig. 5.22 Analytical system [43]

Let the right side of Eqs. 5.39 and 5.40 be zero, then these two equations can be rewritten as

$$d\theta/dt = 0 \tag{5.41}$$

$$d\varphi/dt = 0 \tag{5.42}$$

Then, $\theta\,(t)$ and $\varphi(t)$ become constants. Here we define $\theta(0) = \theta_0$, $\varphi(0) = \varphi_0$.

$$\theta = \theta_0 \tag{5.43}$$

$$\varphi = \varphi_0 \tag{5.44}$$

In other words, when the imposed magnetic field is too weak or the viscosity of the solution is very large, neither $\theta\,(t)$ nor $\varphi(t)$ will change from their initial value since the magnetization torque for rotation of the crystal is not large enough.

(2) $\tau_s \rightarrow 0\,(H(t) \rightarrow \infty$ or $\mu \rightarrow 0)$

In this case, the left sides of Eqs. 5.39 and 5.40 become zero.

$$cos^2(\varphi - \omega t)\ sin2\theta = 0 \tag{5.45}$$

$$sin\ 2(\varphi - \omega t) = 0 \tag{5.46}$$

The solution of Eq. 5.46 is:

$$\varphi(t) = -\omega t = \pi\,(2n + 1)/2,\ \ (n = 0,\ 1,\ 2,\ 3,\ \ldots) \tag{5.47}$$

Therefore, the angle $\varphi(t)$ always keeps a phase difference π/2 radian with the imposed magnetic field and rotates with the same speed of ω. Then, Eq. 5.45 becomes:

$$0 \times sin2\theta = 0 \tag{5.48}$$

Equation 5.45 holds for any value of θ. Therefore, in this case, the torque for rotation in the θ component never generates. That is, when the imposed magnetic field is too large or the viscosity of the solution is too small, the angle θ cannot be changed.

(3) $0 < \tau_s < \infty$

In this range, there are two cases existing: the angle $\varphi(t)$ synchronizing with the rotating magnetic field or not. In the first case, the crystal alignment time can be theoretically derived as follows:

In the case when the horizontal component of the angular speed of a rotating crystal is the same as that of the magnetic field, the relation between $\varphi(t)$ and ω in steady state is

$$d\varphi/dt = \omega \tag{5.49}$$

Then, $\varphi(t)$ can be written as

$$\varphi(t) = \omega t + \delta \tag{5.50}$$

Here, the integral constant δ can be obtained by using Eqs. 5.50 and 5.40,

$$d\delta/dt = d(\varphi - \omega t)/dt = (1/2\tau_s)\ sin\,2\delta - \omega = 0 \tag{5.51}$$

From Eq. 5.51, we get the following solution:

$$\delta = arcsin\,(2\tau_s\omega)/2 = arcsin(12\mu\omega/\Delta\chi\mu_{m0}H^2)/2 \tag{5.52}$$

Hence, the synchronous condition is given as $\tau_S\omega \leqq 0.5$, otherwise it can be step-out. By substituting Eq. 5.50 into Eq. 5.39, the differential equation of $\theta(t)$ is expressed as follows:

$$d\theta/dt = (-1/2\tau_s)\,cos^2\,\delta \cdot sin2\theta \tag{5.53}$$

This equation can be solved by using the initial condition of $\theta = \theta_0$ at $t = 0$ as Eq. 5.54.

$$tan\theta = tan\theta_0 \cdot exp(-cos^2\delta t/\tau_s) = tan\theta_0 \cdot exp(-\omega t/tan\delta) \tag{5.54}$$

It can be found that the angle $\theta(t)$ decreases with elapsing of time except for $\theta_0 = \pi/2$. Moreover, from Eqs. 5.54 and 5.52, the crystal alignment time τ_R in the case where $\varphi(t)$ of the crystal is synchronized with the imposed magnetic field is given as Eq. 5.55.

$$\tau_R = tan\delta/\omega = 2\tau_s \Big/ \left\{1 - \sqrt{1 - 4(\tau_s\omega)^2}\right\} \tag{5.55}$$

As mentioned above, in the synchronized range, the most favorable condition for crystal alignment is $\tau_s\omega = 0.5$, and the crystal alignment time is two times longer than that under a static magnetic field ($\tau_R = 2\tau_s$ is obtained from Eq. 5.55). By differentiating τ_R with respect to magnetic field H, viscosity μ and rotation speed ω, Eqs. 5.56, 5.57, 5.58 can be obtained, respectively.

$$\partial\tau_R/\partial H = \left\{1 + \sqrt{1 - 4(\tau_s\omega)^2}\right\} \Big/ \left\{\tau_s H\omega^2\sqrt{1 - 4(\tau_s\omega)^2}\right\} > 0 \quad (5.56)$$

$$\partial\tau_R/\partial\mu = -\left\{1 + \sqrt{1 - 4(\tau_s\omega)^2}\right\} \Big/ \left\{2\tau_s\mu\omega^2\sqrt{1 - 4(\tau_s\omega)^2}\right\} < 0 \quad (5.57)$$

$$\partial\tau_R/\partial\omega = -2\tau_s \Big/ \left\{\omega\sqrt{1 - 4(\tau_s\omega)^2}\right\} - \left\{1 + \sqrt{1 - 4(\tau_s\omega)^2}\right\} \Big/ (\tau_s\omega)^2 < 0 \quad (5.58)$$

From Eq. 5.56, we can conclude that the alignment time τ_R increases with an increasing magnetic field, and from Eq. 5.57, the alignment time τ_R decreases with the increase of viscosity. That is, contrary to the situation when a static magnetic field is imposed, in the synchronized range, the weaker the imposed magnetic field, as well as the larger the viscosity, the shorter the alignment time. In addition, the crystal aligns much more quickly under a high speed of rotation. The conditions of (1), (2) and (3) used for the theoretical analysis are schematically shown in Fig. 5.23. When the crystal alignment time under a static magnetic field is $0 < \tau_S < \infty$, crystal alignment behavior in the synchronized region has been theoretically clarified. In order to experimentally clarify the relation between magnetic field strength and crystal alignment time, a model experiment has been carried out. The experimental setup is schematically shown in Fig. 5.24. A mechanically rotated stage, on which an alumina column-shaped vessel with a size of ϕ 20 mm x 40 mm was placed, was set in the bore of the magnetic field with its rotating axis perpendicular to the magnetic field direction. A polymer fiber was dispersed in a vessel containing 4 ml of solution with viscosity of 5 Pa·s, which was prepared by mixing distilled water and glycerin. Magnetic fields of 4 T, 5 T and 6 T were imposed on the vessel which was rotated with a rotation speed of 0.3 rad·s^{-1}. In this system, the polymer fiber aligns parallel to the direction of the rotation axis under a time-varying magnetic field. The alignment behavior of the polymer fiber was observed by a video camera. By using a prism set above the vessel, pictures were taken from the upper and side surfaces at the same time. The pictures taken during the experiment are shown in Fig. 5.25 **a–c**. The pictures shown at the lower parts were taken from the side view of the suspended polymer fiber, while the upper pictures were taken from the upper side surface. In the case of 4 T, the alignment completes in 10s with the polymer fiber parallel to the rotation axis. With an increase of the magnetic field, the alignment time increases. In the case of a 6 T magnetic field (Fig. 5.25 **c**), the alignment of the polymer fiber had not competed even after rotation for 20 s.

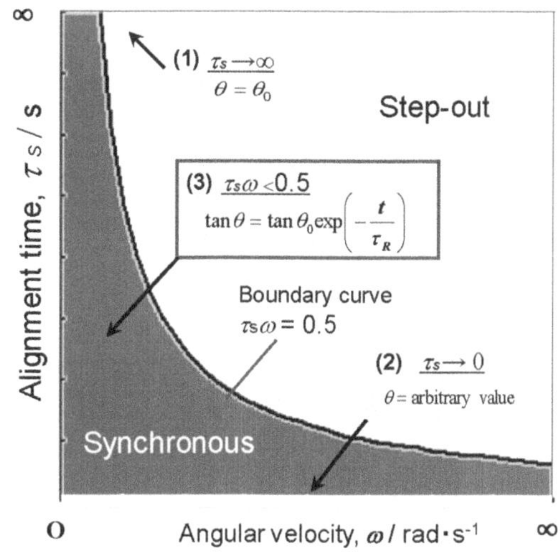

Fig. 5.23 Relation between crystal alignment time τ_S and angular velocity ω [43]

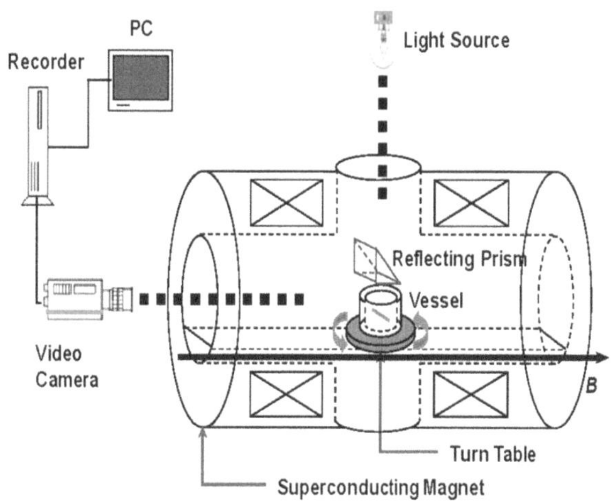

Fig. 5.24 Schematic view of the experimental apparatus [43]

4.4.5 Modulated Magnetic Field [44]

Regarding the rotating magnetic field, a further study was done by Kimura et al. [44]. They found that a rotating magnetic field, whose intensity or frequency is periodically modulated, can produce a 3-dimensional crystalline alignment of a biaxial crystal. A sample prepared by this technique gave rise to an X-ray diffraction pattern nearly equivalent to a real single crystal [45]. Furthermore, this new technology allows us to fabricate a pseudo-single crystal from any crystalline powder with crystal anisotropy. This is expected to open the gate for developing a new field of Magneto-Science.

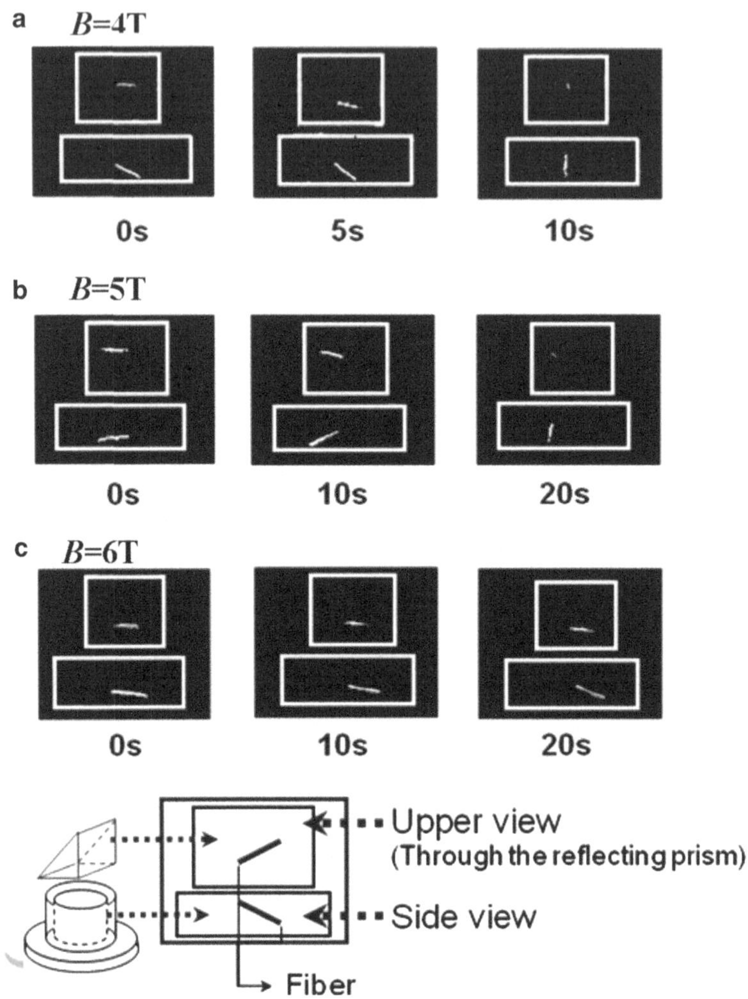

Fig. 5.25 Alignment of polymeric fibers under a high magnetic field and mold rotation. **a** $B = 4$ T, **b** $B = 5$ T, **c** $B = 6$ T [43]

4.5 Sintering Process

Wu and Murakami et al. [46] examined the effect of a magnetic field on crystal alignment in a sintering process. The experimental results are given in the following. Figure 5.26 shows the relative facial angle defined in Appendix [47], which indicates a lumped crystalline angle obtained by using an X-ray diffraction analysis pattern. When the green sample of HAp (Hydroxyapatite) was prepared by using slip casting under no magnetic field and sintered in a 10 T magnetic field, the figure changed

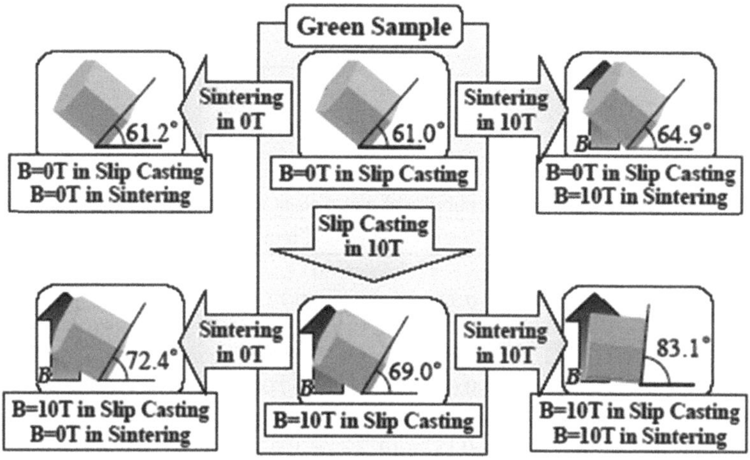

Fig. 5.26 Relative facial angles from c-plane of samples [46]

from 61.0° to 64.9°. These figures imply that the effect of the magnetic field in the sintering is detectable, but little. On the other hand, when the green samples were prepared by using slip casting under a magnetic field, the relative facial angle increased to 72.4° by sintering without a magnetic field, and further highly increased to 83.1° by sintering with a magnetic field. This result is explained as follows: as slip casting with a magnetic field introduces some amount of crystal alignment in the green sample, which could preferentially agglomerate in the sintering process, because the agglomeration resistance of the aligned crystals contacting with the same crystalline plane each other is lower than that of non-aligned crystals. Thus, the aligned crystals more quickly agglomerate than the non-aligned ones. Furthermore, the Ostwald ripening mechanism works to enlarge the agglomerated crystals with a larger size rather than the non-aligned crystals with a smaller size. That is, the double effects of the agglomeration and ripening mechanisms accelerate the crystal alignment in the sample that was treated by the imposition of a magnetic field in the slip casting and sintering processes. This mechanism can be explained in the theoretical prediction given by Yasuda et al. [19].

5 Fluid Motion Driven by a Magnetization Force

There are some reports on such interesting phenomena as that the vaporization rate of water is enhanced by a high magnetic field [4], and the absorption rate of oxygen in water is increased by a high magnetic field [5]. These phenomena are understood as being caused by the magnetization force. In order to compare the fluid motion induced by a gravity force and the fluid motion induced by the magnetization force (hereafter this motion is called a magnetic convection), the comparison between

Table 5.8 Comparison between gravity force and magnetization force

Gravity force	Magnetization force
$\rho \boldsymbol{g}$	$\chi\mu_{m0}(\boldsymbol{H}\cdot\nabla)\boldsymbol{H}$
$\boldsymbol{g}$	$(\boldsymbol{H}\cdot\nabla)\boldsymbol{H}$
ρ	$\chi\mu_{m0} = \chi_g\rho\mu_{m0} \propto \pm\rho$ χ_g: Mass magnetic susceptibility

Table 5.9 Magnetic susceptibility of materials at room temperature [50]

	$\chi''(-)$	$\chi_g''(\text{cm}^3/\text{g})$	ρ'' (g/cm^3)
Vacuum	0	0	0
Water	-0.72×10^{-6}	-0.72×10^{-6}	1.0
Acetone	-0.463×10^{-6}	-0.585×10^{-6}	0.7853
Benzene	-0.518×10^{-6}	-0.702×10^{-6}	0.874
Ethanol	-0.617×10^{-6}	-0.718×10^{-6}	0.786
N_2(gas)	-0.54×10^{-9}	-0.43×10^{-6}	1.25×10^{-3}
O_2(gas)	$+0.152\times10^{-6}$	$+106.2\times10^{-6}$	1.43×10^{-3}
Bismuth	-0.132×10^{-4}	-0.135×10^{-5}	9.78
Diamond	-0.17×10^{-5}	-0.49×10^{-6}	3.51
Graphite (a or b axis)	-0.11×10^{-5}	-0.50×10^{-6}	2.27
Graphite (c-axis)	-0.49×10^{-4}	-0.22×10^{-4}	2.27
Al	$+0.175\times10^{-5}$	$+0.63\times10^{-6}$	2.70
Ag (at 961 C)	$+0.184\times10^{-5}$	$+0.135\times10^{-6}$	13.6

χ'' magnetic susceptibility; χ_g'' mass magnetic susceptibility; ρ'' density

the gravity force of $\rho\boldsymbol{g}$ and the magnetization force of $\chi\mu_{m0}(\boldsymbol{H}\cdot\nabla)\boldsymbol{H}$ is listed in Table 5.8. Here, an acceleration vector of gravity $\boldsymbol{g}$ corresponds to $(\boldsymbol{H}\cdot\nabla)\boldsymbol{H}$ for the magnetization force. We cannot change the value and direction of $\boldsymbol{g}$, but those of $(\boldsymbol{H}\cdot\nabla)\boldsymbol{H}$ can be changed by controlling the magnetic field. On the other hand, the density ρ as a physical property corresponds to $\chi\mu_{m0}$for the magnetization force. Though the density of ρ is always positive, the magnetic susceptibility of χ is positive for a paramagnetic substance and negative for diamagnetic. Moreover, in the non-magnetic (paramagnetic and diamagnetic) substances, mass magnetic susceptibility χ_g'' has almost the same value in many substances, except for oxygen as shown in Table 5.9.

The volumetric magnetic susceptibility χ is given as Eq. 5.59.

$$\chi = \rho\cdot\chi_g,\ (\text{or}\ \chi'' = \rho''\cdot\chi_g'') \tag{5.59}$$

From $\chi = 4\pi\chi''$(see Eq. 5.10), the relation of $\chi\mu_{m0} = 4\pi\rho''\cdot\chi_g''\mu_{m0} \propto \rho'' \propto \rho$ holds. That is, $\chi\mu_{m0}$ is in proportion to ρ. In Chap. 4, it was mentioned that inclusions with electrically poor conductivity can be driven in a metal to the opposite direction to a Lorentz force and eliminated. The force of $(\sigma_p - \sigma_{med})\boldsymbol{J}\times\boldsymbol{B}$ which drives an inclusion in the opposite direction of the Lorentz force, was called the **electromagnetic Archimedes force**. The same thing occurs in the case of the magnetization force instead of the Lorentz force. In this case, the driving force acting on an inclusion is $\{(\chi_p - \chi_{med})\mu_{m0}\}\,(\boldsymbol{H} \cdot \nabla \boldsymbol{H})$. This driving force was first named the "**magneto-Archimedes buoyant force**" [48] and then the "**magnetic Archimedes force**" [49]. By using this force, we are able to move an inclusion close to or far from a solidifying front. In future, we will be able to eliminate inclusions by using the magnetic Archimedes force when a high intensity magnetic field is introduced in production processes.

5.1 *Equation of Motion Taking Account Magnetization Force*

By introducing the magnetization force into the external force term in Eq. 3.2, Eq. 5.60 is obtained.

$$\rho\{\partial \boldsymbol{v}/\partial t + (\boldsymbol{v} \cdot \nabla)\boldsymbol{v}\} = -\nabla p + \mu\nabla^2\boldsymbol{v} + (\chi\mu_{mo})(\boldsymbol{H} \cdot \nabla)\boldsymbol{H} \tag{5.60}$$

The vector operation of rotation $\nabla\times$ to both sides of the above equation yields Eq. 5.61.

$$\partial\boldsymbol{\omega}/\partial t = \nabla \times (\boldsymbol{v} \times \boldsymbol{\omega}) + \nu\nabla^2\boldsymbol{\omega} + (1/\rho)\nabla \times \{(\chi\mu_{mo})(\boldsymbol{H} \cdot \nabla)\boldsymbol{H}\} \tag{5.61}$$

The third term in the right hand side of Eq. 5.61 is the driving force term of magnetic convection. Eq. 5.61 does not indicate a magnetic field, but the gradient of the magnetic field derives a fluid motion.

In order to visualize this theoretical consideration, Wakayama [50] conducted an experiment as follows: nitrogen gas containing water moisture was allowed to flow into the air space between the poles of a magnet as shown in Fig. 5.27. The imposition of a magnetic field forces the nitrogen gas to be a jet flow in the direction reducing magnetic field intensity. The stream lines were obtained by solving the Navier-Stokes equation, taking into account the magnetization force. It is understood that a vortex is produced by imposition of a magnetic field and this calculated result, shown in the right-hand sides of **a** and **b** in Fig. 5.27, qualitatively agrees with the experimental result.

5.2 *Dimensionless Numbers Relating Magnetization Force*

New dimensionless numbers are obtained by substituting the magnetization force instead of the gravity force into dimensionless numbers related with the gravity force.

$$Fr_m = (\rho V^2/L)/\{(\chi/\mu_m)(B^2/L_m)\} = \rho\mu_m V^2 L_m/\chi\, B^2 L$$

magnetic Froude number

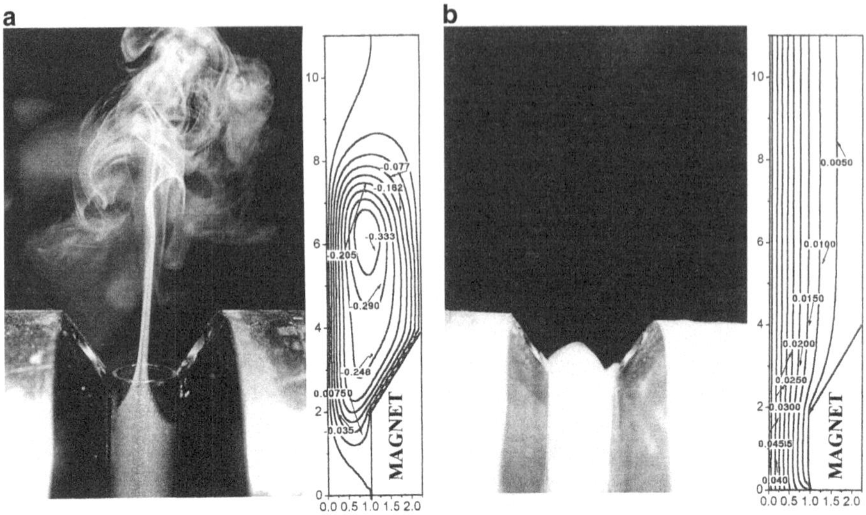

Fig. 5.27 The visualization of the nitrogen gas flow spouted into the air and the calculated result of stream lines. **a** with a magnetic field and **b** without magnetic field [50]

$$Gr_m = \{(\chi/\mu_m)(B^2/L_m)(\rho V^2/L)\}/\{\mu^2(1/L^2)^2V^2\} = L^3\ \rho\chi B^2/L_m\mu^2\mu_m$$

magnetic Grashof number

Here, L_m is the characteristic length related with a magnetic gradient.

$$Ra_m = Gr_m \cdot Pr$$ *magnetic Rayleigh number*

Also, the following dimensionless number is obtained from the ratio of the magnetization force to the gravity force.

$$A = \{(\chi/\mu_m)(B^2/L_m)\}/\{\rho g\} = \chi B^2/(L_m\mu_m\rho g)$$

magnetic gravity ratio number

When $A = 1$, a material can be levitated in air due to the magnetization force. When a material with χ_p and ρ_p exists in a medium with χ_{med} and ρ_{med}, the magnetic gravity ratio number is modified as follows:

$$A' = (\chi_p - \chi_{med})B^2/\{(\rho_p - \rho_{med})L_m\mu_m g\}$$

relative magnetic gravity ratio number

A relative magnetic gravity ratio number also can be derived from Eq. 5.5 expressing the enhanced Moses effect. These days, people pay close attention to the creation of new materials under micro-gravity making use of an enclosed space simulating outer space. In this case, the required point is mostly not micro-gravity, but elimination of contamination from a vessel. Micro-gravity does not provide a suitable atmosphere for materials processing, because the result might be pores in products due to lack of bubble elimination and production of non-homogeneous materials due to lack of mixing. Regarding this point, levitation using the magnetization force can make a material float even under gravitation. That is, a pseudo space can be created on Earth by making use of the magnetization force. This method not only is economical in comparison with outer space, but also leads to new methods for materials processing.

6 Future Prospect

Application of an electromagnetic force to materials processing, so-called "Electromagnetic Processing of Materials (EPM)" has been recognized as a cutting edge technology, especially in the fields of advanced materials processing. Recently, due largely to advances in super-conducting magnets, high magnetic fields have become readily available and are being applied in various fields of science. The result has been discovery of many interesting phenomena relating to the high magnetic field

Fig. 5.28 A tree of Electromagnetic Processing of Materials and a sprout of materials processing by use of a high magnetic field

and emergence of a new academic area. This area, which we will metaphorically refer to as "**Science Relating with a High Magnetic Field**", can be expected to open a gate for new research. In order to connect the seeds sprouting from "Science Relating with a High Magnetic Field" with the needs of "Materials Science and Engineering", the new area of "**Materials Processing by Use of a High Magnetic Field**" is expected to grow under the umbrella of "**Electromagnetic Processing of Materials**". This is illustrated as a strip cartoon in Fig. 5.28.

Appendix [47]

The orientation index of $(h_i,\ k_i, l_i)$ plane N_{h_i,k_i,l_i} defined as Eq. 5.62, is evaluated from the X-ray diffraction patterns.

$$N_{h_i,k_i,l_i} = F_{h_i,k_i,l_i} / F^0_{h_i,k_i,l_i}, \tag{5.62}$$

where F_{h_i,k_i,l_i} is an intensity fraction of a $(h_i,\ k_i,\ l_i)$ plane and defined by Eq. 5.63 and $F^0_{h_i,k_i,l_i}$ is obtained from the standard data of JCPDS cards.

$$F_{h_i,k_i,l_i} = I_{h_i,k_i,l_i} / (I_{h_1,k_1,l_1} + I_{h_2,k_2,l_2} + I_{h_3,k_3,l_3} + \cdots\cdots + I_{h_n,k_n,l_n}), \tag{5.63}$$

where I_{h_i,k_i,l_i} is intensity for the diffraction line of $(h_i,\ k_i,\ l_i)$.

Moreover, in order to comprehensively evaluate the over-all degree of crystalline texture, the definition of a relative facial angle θ_F is obtained by Eq. 5.64.

$$\theta_F = \sum (I_{h_i,k_i,l_i} \times \theta_{h_i,k_i,l_i}) \Big/ \sum I_{h_i,k_i,l_i} \tag{5.64}$$

θ_{h_i,k_i,l_i} is the facial angle between (h_i, k_i, l_i) and $(0, 0, n)$ planes. The relative facial angle θ_F is reduced to 0° when all crystals are oriented to the $(0, 0, n)$ plane and to 90° when oriented to the plane perpendicular to $(0, 0, n)$.

References

1. P. de Rango, M. Lees, P. Lejay, A. Sulpice, R. Tournier, M. Ingold, P. Germi, M. Pernet, Nature **349**, 770 (1991)
2. N. Hirota, T. Honma, H. Sugawara, K. Kitazawa, M. Iwasaka, S. Ueno, H. Yokoi, Y. Kakudate, S. Fujiwara, M. Kawamura, Jpn. J. Appl. Phys. **34**, L991 (1995)
3. M.V. Berry, A.K. Geim, Eur. J. Phys. **18**, 307 (1997)
4. J. Nakagawa, N. Hirota, K. Kitazawa, M. Syoda, in *1st Symposium on New Magnetic-science Program & Abstracts*, Nov 1997, p. 226
5. Y. Ikezoe, N. Hirota, T. Sakihama, K. Mogi, H. Uetake, J. Nakagawa, K. Sugawara, K. Kitazawa, in *1st Symposium on New Magnetic-Science Program & Abstracts*, Nov 1997, p. 231
6. S. Asai, J. Jpn. Inst. Met. **61**, 1271 (1997)
7. E. Beaugnon, T. Tournier, Nature **349**, 470 (1991)

8. A. Lusnikov, L.L. Miller, R.W. McCaullum, S. Mitra, W.C. Lee, D.C. Johnson, J. Appl. Phys. **65**, 3136 (1989)
9. J.E. Tkazyk, K.W. Lay, J. Mater. Res. **5**, 1368 (1990)
10. P. de Rango, M. Lee, P. Lejay, A. Sulpice, R. Tournier, M. Ingold, P. Gerni, M. Pernet, Nature **349**, 770 (1991)
11. R.H. Arendt, M.F. Garbauskas, K.W. Lay, J.E. Tkaczyk, Physica. C **176**, 131 (1991)
12. A. Holloway, R.W. McCallun, S.R. Arrasmith, J. Mater. Res. **8**, 727 (1993)
13. S. Stassenn, R. Cloots, A. Rulmont, F. Gillet, H. Bougrine, P.A. Godelaine, A. Dang, M. Ausloos, Physica. C **235–240**, 515 (1994)
14. S. Stassenn, R. Cloots, Rh Vanderbemden, P.A. Godelaine, H. Bougrine, A. Rulmont, M. Ausloos, J. Mater. Res. **11**, 1082 (1996)
15. N. Hirota, T. Hon-ma, Kin-zoku **65**(9), 793 (1995)
16. M. Matsui, *MSJ Summer School, The Basis of Applied Magnetic* (MSJ, 1995, 1996), p. 1
17. K. Ohta, *Ziki Kougaku no Kiso* (Kyoritsu Shuppan, Tokyo 1993), p. 42
18. A.E. Mikelson, Kh Karkin, J Cryst. Growth **52**, 524 (1981)
19. H. Yasuda, K. Tokieda, I. Ohnaka, Mater. Trans. JIM **41**, 1005 (2000)
20. B.A. Legrand, R. Perrier de la Bathie, R. Tournier, in *Proceedings of Int. Cong. of Electromagnetic Processing of Materials*, vol. 2, Paris, May 1997, p. 309
21. P. Courtois, R. Perrier de la Bathie, R. Tournier, in *Proceedings of Int. Cong. of Electromagnetic Processing of Materials,* vol. 2, Paris, May 1997, p. 277
22. T. Sugiyama, M. Tahashi, K. Sassa, S. Asai, Trans. ISIJ **43**, 855 (2003)
23. T. Taniguchi, K. Sassa, T. Yamada, S. Asai, Mater. Trans. JIM **41**, 981 (2000)
24. M. Tahashi, K. Sassa, I. Hirabayashi, S. Asai, Mater. Trans. Jim **41**, 985 (2000)
25. Y. Lu, A. Nagata, K. Watanabe, T. Nojima, K. Sugawara, S. Hanada, S. Kamada, Physica C **392**, 453 (2003)
26. S.S. He, Y.D.D. Zhang, X. Zhao, L. Zuo, J.C.C. He, K. Watanabe, T. Zhang, G. Nishijima, C. Esling, Adv. Eng. Mater. **5**, 579 (2003)
27. P. Chen, H. Maeda, K. Watanabe, M. Motokawa, Physica C **337**, 160 (2000)
28. P. Chen, H. Maeda, K. Watanabe, M. Motokawa, H. Kitaguchi, H. Kumakura, Physica C **324**, 172 (1999)
29. P. Chen, H. Maeda, K. Kakimoto, P.X. Zhang, K. Watanabe, M. Motokawa, Physica C **320**, 96 (1999)
30. M.H. Zimmerman, K.T. Faber, E.R. Fuller Jr., J. Am. Ceram. Soc. **80**, 2725 (1997)
31. E. Farrel, B.S. Chandrasekhar, M.R. DeGuire, M.M. Fang, V.G. Kogan, J.R. Clem, D.K. Finnemore, Phys. Rev. B **36**, 4025 (1987)
32. M. Ferreira, M.B. Maple, H. Zhou, R.R. Hake, B.W. Lee, C.L. Seaman, M.V. Kuric, R.P. Guertin, Appl. Phys. A **7**, 105 (1988)
33. W. Paulik, K.T. Faber, E.R. Fullar Jr., J. Am. Ceram. Soc. **77**, 454 (1994)
34. K. Inoue, K. Sassa, Y. Yokogawa, Y. Sakka, M. Okido, S. Asai, Mater. Trans. JIM **44**, 1133 (2003)
35. Y. Sakka, T.S. Suzuki, N. Tanabe, S. Asai, K. Kitazawa, Jpn. J. Appl. Phys. **41**, 1416 (2002)
36. M. Mizushima, J. Okada, *Tanso Zairyou* (Kyoritsu Shuppan, Tokyo 1970), p. 157
37. C. Wu, S. Li, K. Sassa, Y. Chino, K. Hattori, S. Asai, Mater. Trans. **46**, 1311 (2005)
38. T.S. Suzuki, H. Otsuka, Y. Sakka, K. Hiraga, K. Kitazawa, J. Jpn. Soc. Powder & Powder Metallurgy **47**, 1010 (2000)
39. T.S. Suzuki, Y. Sakka, K. Kitazawa, Adv. Eng. Mater. **3**, 490 (2001)
40. T. Kimura, Polym. J. **35**, 823 (2003)
41. S. Li, K. Sassa, K. Iwai, S. Asai, Mater. Trans. **45**, 3124 (2004)
42. T. Kimura, M. Yamato, W. Koshimizu, M. Koike, T. Kawai, Langmuir **16**, 858 (2000)
43. J. Akiyama, H. Asano, K. Iwai, S. Asai, J. Jpn. Inst. Met. **71**, 108 (2007)
44. T. Kimura, M. Yoshino, Langmuir **21**, 4805 (2005)
45. T. Kimura, F. Kimura, M. Yoshino, Langmuir **22**, 3464 (2006)
46. C. Wu, Y. Murakami, K. Sassa, K. Iwai, S. Asai, Key Eng. Mater. **75**, 284 (2005)
47. M. Tahashi, M. Ishihara, K. Sassa, S. Asai, Mater. Trans. JIM **44**, 285 (2003)

48. Y. Ikezoe, N. Hirota, J. Nakagawa, K. Kitazawa, Nature **393**, 749 (1998)
49. S. Asai,*Zairyou Denzi Purossesingu* (Uchida Rokaku Ho, Tokyo, 2000), p. 103
50. N. Wakayama, J. Jpn. Inst. Met. **61**, 1272 (1997)

Appendixes 1–8

Appendix 1: The Equation of Continuity for Incompressible Fluid

Rectangular (Cartesian) coordinates (x, y, z)

$$\frac{\partial}{\partial x}(v_x) + \frac{\partial}{\partial y}(v_y) + \frac{\partial}{\partial z}(v_z) = 0 \tag{A.1}$$

Circular cylindrical coordinates (r, θ, z)

$$\frac{1}{r}\frac{\partial}{\partial r}(rv_r) + \frac{1}{r}\frac{\partial}{\partial \theta}(v_\theta) + \frac{\partial}{\partial z}(v_z) = 0 \tag{A.2}$$

Spherical coordinates (r, θ, φ)

$$\frac{1}{r^2}\frac{\partial}{\partial r}\left(r^2 v_r\right) + \frac{1}{r\,sin\,\theta}\frac{\partial}{\partial \theta}(v_\theta\, sin\theta) + \frac{1}{r\,sin\,\theta}\frac{\partial}{\partial \varphi}(v_\varphi) = 0 \tag{A.3}$$

Appendix 2: The Conservation of Mass for Species

Rectangular (Cartesian) coordinates (x, y, z)

$$\frac{\partial c}{\partial t} + v_x\frac{\partial c}{\partial x} + v_y\frac{\partial c}{\partial y} + v_z\frac{\partial c}{\partial z} = D\left(\frac{\partial^2 c}{\partial x^2} + \frac{\partial^2 c}{\partial y^2} + \frac{\partial^2 c}{\partial z^2}\right) \tag{A.4}$$

Circular cylindrical coordinates (r, θ, z)

$$\frac{\partial c}{\partial t} + v_r\frac{\partial c}{\partial r} + v_\theta\frac{1}{r}\frac{\partial c}{\partial \theta} + v_z\frac{\partial c}{\partial z} = D\left\{\frac{1}{r}\frac{\partial}{\partial r}\left(r\frac{\partial c}{\partial r}\right) + \frac{1}{r^2}\frac{\partial^2 c}{\partial \theta^2} + \frac{\partial^2 c}{\partial z^2}\right\} \tag{A.5}$$

S. Asai, *Electromagnetic Processing of Materials*, Fluid Mechanics and Its Applications 99, DOI 10.1007/978-94-007-2645-1,
© Springer Science+Business Media B.V. 2012

Spherical coordinates (r, θ, φ)

$$\frac{\partial c}{\partial t} + v_r \frac{\partial c}{\partial r} + v_\theta \frac{1}{r}\frac{\partial c}{\partial \theta} + v_\varphi \frac{1}{r \sin\theta}\frac{\partial c}{\partial \varphi} = D\left\{\frac{1}{r^2}\frac{\partial}{\partial r}\left(r^2 \frac{\partial c}{\partial r}\right) + \frac{1}{r^2 \sin\theta}\frac{\partial}{\partial \theta}\left(\sin\theta \frac{\partial c}{\partial \theta}\right) + \frac{1}{r^2 \sin^2\theta}\frac{\partial^2 c}{\partial \varphi^2}\right\} \tag{A.6}$$

Appendix 3: The Equation of Energy for Incompressible Fluid

Rectangular (Cartesian) coordinates (x, y, z)

$$\rho c_p\left(\frac{\partial T}{\partial t} + v_x \frac{\partial T}{\partial x} + v_y \frac{\partial T}{\partial y} + v_z \frac{\partial T}{\partial z}\right) = \lambda\left(\frac{\partial^2 T}{\partial x^2} + \frac{\partial^2 T}{\partial y^2} + \frac{\partial^2 T}{\partial z^2}\right) \tag{A.7}$$

Circular cylindrical coordinates (r, θ, z)

$$\rho c_p\left(\frac{\partial T}{\partial t} + v_r \frac{\partial T}{\partial r} + v_\theta \frac{1}{r}\frac{\partial T}{\partial \theta} + v_z \frac{\partial T}{\partial z}\right) = \lambda\left\{\frac{1}{r}\frac{\partial}{\partial r}\left(r\frac{\partial T}{\partial r}\right) + \frac{1}{r^2}\frac{\partial^2 T}{\partial \theta^2} + \frac{\partial^2 T}{\partial z^2}\right\} \tag{A.8}$$

Spherical coordinates (r,θ, φ)

$$\rho c_p\left(\frac{\partial T}{\partial t} + v_r \frac{\partial T}{\partial r} + v_\theta \frac{1}{r}\frac{\partial T}{\partial \theta} + v_\varphi \frac{1}{r \sin\theta}\frac{\partial T}{\partial \varphi}\right) = \lambda\left\{\frac{1}{r^2}\frac{\partial}{\partial r}\left(r^2 \frac{\partial T}{\partial r}\right) + \frac{1}{r^2 \sin\theta}\frac{\partial}{\partial \theta}\left(\sin\theta \frac{\partial T}{\partial \theta}\right) + \frac{1}{r^2 \sin^2\theta}\frac{\partial^2 T}{\partial \varphi^2}\right\} \tag{A.9}$$

Appendix 4: The Components of the Energy Flux *q*

Rectangular (Cartesian) coordinates (x, y, z)

$$q_x = -\lambda \frac{\partial T}{\partial x} \tag{A.10}$$

$$q_y = -\lambda \frac{\partial T}{\partial y} \tag{A.11}$$

$$q_z = -\lambda \frac{\partial T}{\partial z} \tag{A.12}$$

Circular cylindrical coordinates (r, θ, z)

$$q_r = -\lambda \frac{\partial T}{\partial r} \tag{A.13}$$

$$q_\theta = -\lambda \frac{1}{r} \frac{\partial T}{\partial \theta} \tag{A.14}$$

$$q_z = -\lambda \frac{\partial T}{\partial z} \tag{A.15}$$

Spherical coordinates (r, θ, φ)

$$q_r = -\lambda \frac{\partial T}{\partial r} \tag{A.16}$$

$$q_\theta = -\lambda \frac{1}{r} \frac{\partial T}{\partial \theta} \tag{A.17}$$

$$q_\varphi = -\lambda \frac{1}{r \sin\theta} \frac{\partial T}{\partial \varphi} \tag{A.18}$$

Appendix 5: The Equation of Motion for a Newtonian Fluid with Constant ρ and μ

Rectangular (Cartesian) coordinates (x, y, z)
(x-component)

$$\rho\left(\frac{\partial v_x}{\partial t} + v_x \frac{\partial v_x}{\partial x} + v_y \frac{\partial v_x}{\partial y} + v_z \frac{\partial v_x}{\partial z}\right) = -\frac{\partial p}{\partial x} + \mu\left(\frac{\partial^2 v_x}{\partial x^2} + \frac{\partial^2 v_x}{\partial y^2} + \frac{\partial^2 v_x}{\partial z^2}\right) + f_x \tag{A.19}$$

(y-component)

$$\rho\left(\frac{\partial v_y}{\partial t} + v_x \frac{\partial v_y}{\partial x} + v_y \frac{\partial v_y}{\partial y} + v_z \frac{\partial v_y}{\partial z}\right) = -\frac{\partial p}{\partial y} + \mu\left(\frac{\partial^2 v_y}{\partial x^2} + \frac{\partial^2 v_y}{\partial y^2} + \frac{\partial^2 v_y}{\partial z^2}\right) + f_y \tag{A.20}$$

(z-component)

$$\rho\left(\frac{\partial v_z}{\partial t} + v_x \frac{\partial v_z}{\partial x} + v_y \frac{\partial v_z}{\partial y} + v_z \frac{\partial v_z}{\partial z}\right) = -\frac{\partial p}{\partial z} + \mu\left(\frac{\partial^2 v_z}{\partial x^2} + \frac{\partial^2 v_z}{\partial y^2} + \frac{\partial^2 v_z}{\partial z^2}\right) + f_z \tag{A.21}$$

Circular cylindrical coordinates (r, θ, z)
(r-component)

$$\rho\left(\frac{\partial v_r}{\partial t}+v_r\frac{\partial v_r}{\partial r}+\frac{v_\theta}{r}\frac{\partial v_r}{\partial \theta}-\frac{v_\theta^2}{r}+v_z\frac{\partial v_r}{\partial z}\right)=-\frac{\partial p}{\partial r}$$
$$+\mu\left[\frac{\partial}{\partial r}\left\{\frac{1}{r}\frac{\partial(rv_r)}{\partial r}\right\}+\frac{1}{r^2}\frac{\partial^2 v_r}{\partial\theta^2}-\frac{2}{r^2}\frac{\partial v_\theta}{\partial\theta}+\frac{\partial^2 v_r}{\partial z^2}\right]+f_r \tag{A.22}$$

(θ-component)

$$\rho\left(\frac{\partial v_\theta}{\partial t}+v_r\frac{\partial v_\theta}{\partial r}+\frac{v_\theta}{r}\frac{\partial v_\theta}{\partial \theta}+\frac{v_r v_\theta}{r}+v_z\frac{\partial v_\theta}{\partial z}\right)=-\frac{1}{r}\frac{\partial p}{\partial \theta}$$
$$+\mu\left[\frac{\partial}{\partial r}\left\{\frac{1}{r}\frac{\partial(rv_\theta)}{\partial r}\right\}+\frac{1}{r^2}\frac{\partial^2 v_\theta}{\partial\theta^2}+\frac{2}{r^2}\frac{\partial v_r}{\partial\theta}+\frac{\partial^2 v_\theta}{\partial z^2}\right]+f_\theta \tag{A.23}$$

(z-component)

$$\rho\left(\frac{\partial v_z}{\partial t}+v_r\frac{\partial v_z}{\partial r}+\frac{v_\theta}{r}\frac{\partial v_z}{\partial \theta}+v_z\frac{\partial v_z}{\partial z}\right)=-\frac{\partial p}{\partial z}$$
$$+\mu\left\{\frac{1}{r}\frac{\partial}{\partial r}\left(r\frac{\partial v_z}{\partial r}\right)+\frac{1}{r^2}\frac{\partial^2 v_z}{\partial\theta^2}+\frac{\partial^2 v_z}{\partial z^2}\right\}+f_z \tag{A.24}$$

Spherical coordinates (r, θ, φ)
(r-component)

$$\rho\left(\frac{\partial v_r}{\partial t}+v_r\frac{\partial v_r}{\partial r}+\frac{v_\theta}{r}\frac{\partial v_r}{\partial \theta}+\frac{v_\varphi}{r\,sin\,\theta}\frac{\partial v_r}{\partial\varphi}-\frac{v_\theta^2+v_\varphi^2}{r}\right)=-\frac{\partial p}{\partial r}$$
$$+\mu\left\{\frac{1}{r^2}\frac{\partial^2}{\partial r^2}(r^2v_r)+\frac{1}{r^2 sin\theta}\frac{\partial}{\partial\theta}\left(sin\,\theta\frac{\partial v_r}{\partial\theta}\right)+\frac{1}{r^2 sin^2\theta}\frac{\partial^2 v_r}{\partial\varphi^2}\right\}+f_r \tag{A.25}$$

(θ-component)

$$\rho\left(\frac{\partial v_\theta}{\partial t}+v_r\frac{\partial v_\theta}{\partial r}+\frac{v_\theta}{r}\frac{\partial v_\theta}{\partial \theta}+\frac{v_\varphi}{r\,sin\,\theta}\frac{\partial v_\theta}{\partial\varphi}+\frac{v_r v_\theta}{r}-\frac{v_\varphi^2\,cot\,\theta}{r}\right)=-\frac{1}{r}\frac{\partial p}{\partial\theta}$$
$$+\mu\left[\frac{1}{r^2}\frac{\partial}{\partial r}\left(r^2\frac{\partial v_\theta}{\partial r}\right)+\frac{1}{r^2}\frac{\partial}{\partial\theta}\left\{\frac{1}{sin\,\theta}\frac{\partial(v_\theta\,sin\,\theta)}{\partial\theta}\right\}\right.$$
$$\left.+\frac{1}{r^2\,sin^2\,\theta}\frac{\partial^2 v_\theta}{\partial\varphi^2}+\frac{2}{r^2}\frac{\partial v_r}{\partial\theta}-\frac{2\,cot\,\theta}{r^2\,sin\,\theta}\frac{\partial v_\varphi}{\partial\varphi}\right]+f_\theta \tag{A.26}$$

(φ-component)

$$\rho\left(\frac{\partial v_\varphi}{\partial t}+v_r\frac{\partial v_\varphi}{\partial r}+\frac{v_\theta}{r}\frac{\partial v_\varphi}{\partial \theta}+\frac{v_\varphi}{r\,sin\,\theta}\frac{\partial v_\varphi}{\partial \varphi}+\frac{v_r v_\varphi}{r}+\frac{v_\theta\, v_\varphi}{r}cot\;\theta\right)=-\frac{1}{r\,sin\,\theta}\frac{\partial p}{\partial \varphi}$$
$$+\mu\left[\frac{1}{r^2}\frac{\partial}{\partial r}\left(r^2\frac{\partial v_\varphi}{\partial r}\right)+\frac{1}{r^2}\frac{\partial}{\partial \theta}\left\{\frac{1}{sin\,\theta}\frac{\partial(v_\varphi\, sin\,\theta)}{\partial \theta}\right\}\right.$$
$$\left.+\frac{1}{r^2\,sin^2\,\theta}\frac{\partial^2 v_\varphi}{\partial \varphi^2}+\frac{2}{r^2 sin\theta}\frac{\partial v_r}{\partial \varphi}+\frac{2\,cot\,\theta}{r^2\,sin\,\theta}\frac{\partial v_\theta}{\partial \varphi}\right]+f_\varphi \tag{A.27}$$

Appendix 6: Differential Operation for Scalars and Vectors in Rectangular (Cartesian), Cylindrical and Spherical Coordinates

Rectangular (Cartesian) coordinates (x, y, z)

$$\nabla f=\frac{\partial f}{\partial x}\boldsymbol{i}_x+\frac{\partial f}{\partial y}\boldsymbol{i}_y+\frac{\partial f}{\partial z}\boldsymbol{i}_z \tag{A.28}$$

$$\nabla\cdot\boldsymbol{A}=\frac{\partial A_x}{\partial x}+\frac{\partial A_y}{\partial y}+\frac{\partial A_z}{\partial z} \tag{A.29}$$

$$\nabla\times\boldsymbol{A}=\boldsymbol{i}_x\left(\frac{\partial A_z}{\partial y}-\frac{\partial A_y}{\partial z}\right)+\boldsymbol{i}_y\left(\frac{\partial A_x}{\partial z}-\frac{\partial A_z}{\partial x}\right)+\boldsymbol{i}_z\left(\frac{\partial A_y}{\partial x}-\frac{\partial A_x}{\partial y}\right) \tag{A.30}$$

$$\nabla^2 f=\frac{\partial^2 f}{\partial x^2}+\frac{\partial^2 f}{\partial y^2}+\frac{\partial^2 f}{\partial z^2} \tag{A.31}$$

$$\nabla^2\boldsymbol{A}=\boldsymbol{i}_x\nabla^2 A_x+\boldsymbol{i}_y\nabla^2 A_y+\boldsymbol{i}_z\nabla^2 A_z \tag{A.32}$$

Circular cylindrical coordinates (r,θ, z)

$$\nabla f=\frac{\partial f}{\partial r}\boldsymbol{i}_r+\frac{1}{r}\frac{\partial f}{\partial \theta}\boldsymbol{i}_\theta+\frac{\partial f}{\partial z}\boldsymbol{i}_z \tag{A.33}$$

$$\nabla\cdot\boldsymbol{A}=\frac{1}{r}\frac{\partial}{\partial r}(rA_r)+\frac{1}{r}\frac{\partial A_\theta}{\partial \theta}+\frac{\partial A_z}{\partial z} \tag{A.34}$$

$$\nabla\times\boldsymbol{A}=\boldsymbol{i}_r\left(\frac{1}{r}\frac{\partial A_z}{\partial \theta}-\frac{\partial A_\theta}{\partial z}\right)+\boldsymbol{i}_\theta\left(\frac{\partial A_r}{\partial z}-\frac{\partial A_z}{\partial r}\right)+\boldsymbol{i}_z\frac{1}{r}\left\{\frac{\partial\,(rA_\theta)}{\partial r}-\frac{\partial A_r}{\partial \theta}\right\} \tag{A.35}$$

$$\nabla^2 f = \frac{1}{r}\frac{\partial}{\partial r}\left(r\frac{\partial f}{\partial r}\right) + \frac{1}{r^2}\frac{\partial^2 f}{\partial \theta^2} + \frac{\partial^2 f}{\partial z^2} \tag{A.36}$$

$$[\nabla^2 \boldsymbol{A}]_{\mathrm{r}} = \frac{\partial}{\partial r}\left\{\frac{1}{r}\frac{\partial(rA_r)}{\partial r}\right\} + \frac{1}{r^2}\frac{\partial A_r}{\partial \theta^2} + \frac{\partial^2 A_r}{\partial z^2} - \frac{2}{r^2}\frac{\partial A_\theta}{\partial \theta} \tag{A.37}$$

$$[\nabla^2 \boldsymbol{A}]_\theta = \frac{\partial}{\partial r}\left\{\frac{1}{r}\frac{\partial(rA_\theta)}{\partial r}\right\} + \frac{1}{r^2}\frac{\partial^2 A_\theta}{\partial \theta^2} + \frac{\partial^2 A_\theta}{\partial z^2} + \frac{2}{r^2}\frac{\partial A_r}{\partial \theta} \tag{A.38}$$

$$[\nabla^2 \boldsymbol{A}]_{\mathrm{z}} = \frac{1}{r}\frac{\partial}{\partial r}\left(r\frac{\partial A_z}{\partial r}\right) + \frac{1}{r^2}\frac{\partial^2 A_z}{\partial \theta^2} + \frac{\partial^2 A_z}{\partial z^2} \tag{A.39}$$

Spherical coordinates (r, θ, φ)

$$\nabla f = \frac{\partial f}{\partial r}\boldsymbol{i}_r + \frac{1}{r}\frac{\partial f}{\partial \theta}\boldsymbol{i}_\theta + \frac{1}{r\,sin\,\theta}\frac{\partial f}{\partial \varphi}\boldsymbol{i}_\varphi \tag{A.40}$$

$$\nabla \cdot \boldsymbol{A} = \frac{1}{r^2}\frac{\partial}{\partial r}(r^2 A_r) + \frac{1}{r\,sin\,\theta}\frac{\partial(sin\,\theta A_\theta)}{\partial \theta} + \frac{1}{r\,sin\,\theta}\frac{\partial A_\varphi}{\partial \varphi} \tag{A.41}$$

$$\nabla \times \boldsymbol{A} = \boldsymbol{i}_r\frac{1}{r\,sin\,\theta}\left\{\frac{\partial\left(sin\theta A_\varphi\right)}{\partial \theta} - \frac{\partial A_\theta}{\partial \varphi}\right\} + \boldsymbol{i}_\theta\frac{1}{r}\left\{\frac{1}{sin\,\theta}\frac{\partial A_r}{\partial \varphi} - \frac{\partial(rA_\varphi)}{\partial r}\right\}$$
$$+ \boldsymbol{i}_\varphi\frac{1}{r}\left\{\frac{\partial(rA_\theta)}{\partial r} - \frac{\partial A_r}{\partial \theta}\right\} \tag{A.42}$$

$$\nabla^2 f = \frac{1}{r^2}\frac{\partial}{\partial r}\left(r^2\frac{\partial f}{\partial r}\right) + \frac{1}{r^2 sin\,\theta}\frac{\partial}{\partial \theta}\left(sin\,\theta\frac{\partial f}{\partial \theta}\right) + \frac{1}{r^2 sin^2\,\theta}\frac{\partial^2 f}{\partial \varphi^2} \tag{A.43}$$

$$[\nabla^2 \boldsymbol{A}]_{\mathrm{r}} = \frac{\partial}{\partial r}\left\{\frac{1}{r^2}\frac{\partial}{\partial r}(r^2 A_r)\right\} + \frac{1}{r^2\,sin\,\theta}\frac{\partial}{\partial \theta}\left(sin\,\theta\frac{\partial \mathrm{A_r}}{\partial \theta}\right) + \frac{1}{r^2 sin^2\,\theta}\frac{\partial^2 Ar}{\partial \varphi^2}$$
$$- \frac{2}{r^2\,sin\,\theta}\frac{\partial\left(A_\theta\,sin\,\theta\right)}{\partial \theta} - \frac{2}{r^2\,sin\,\theta}\frac{\partial A_\varphi}{\partial \varphi} \tag{A.44}$$

$$[\nabla^2 \boldsymbol{A}]_\theta = \frac{1}{r^2}\frac{\partial}{\partial r}\left(r^2\frac{\partial A_\theta}{\partial r}\right) + \frac{1}{r^2}\frac{\partial}{\partial \theta}\left\{\frac{1}{sin\,\theta}\frac{\partial}{\partial \theta}(A_\theta\,sin\,\theta)\right\} + \frac{1}{r^2 sin^2\,\theta}\frac{\partial^2 A_\theta}{\partial \varphi^2}$$
$$+ \frac{2}{r^2}\frac{\partial A_r}{\partial \theta} - \frac{2\,cot\,\theta}{r^2\,sin\,\theta}\frac{\partial A_\varphi}{\partial \varphi} \tag{A.45}$$

$$[\nabla^2 \boldsymbol{A}]_\varphi = \frac{1}{r^2}\frac{\partial}{\partial r}\left(r^2\frac{\partial A_\varphi}{\partial r}\right) + \frac{1}{r^2}\frac{\partial}{\partial \theta}\left\{\frac{1}{sin\,\theta}\frac{\partial}{\partial \theta}\left(A_\varphi\,sin\,\theta\right)\right\} + \frac{1}{r^2 sin^2\,\theta}\frac{\partial^2 A_\varphi}{\partial \varphi^2}$$
$$+ \frac{2}{r^2 sin\theta}\frac{\partial A_r}{\partial \varphi} + \frac{2\,cot\,\theta}{r^2\,sin\,\theta}\frac{\partial A_\theta}{\partial \varphi} \tag{A.46}$$

Appendix 7: Vector Identities

$$(\mathbf{A} \times \mathbf{B}) \cdot \mathbf{C} = \mathbf{A} \cdot (\mathbf{B} \times \mathbf{C}) = (\mathbf{C} \times \mathbf{A}) \cdot \mathbf{B} \tag{A.47}$$

$$\mathbf{A} \times (\mathbf{B} \times \mathbf{C}) = \mathbf{B}(\mathbf{A} \cdot \mathbf{C}) - \mathbf{C}(\mathbf{A} \cdot \mathbf{B}) \tag{A.48}$$

$$\nabla \cdot (\nabla \times \mathbf{A}) = 0 \tag{A.49}$$

$$\nabla \times (\nabla f) = 0 \tag{A.50}$$

$$\nabla (fg) = f \nabla g + g \nabla f \tag{A.51}$$

$$\nabla (\mathbf{A} \cdot \mathbf{B}) = (\mathbf{A} \cdot \nabla)\mathbf{B} + (\mathbf{B} \cdot \nabla)\mathbf{A} + \mathbf{A} \times (\nabla \times \mathbf{B}) + \mathbf{B} \times (\nabla \times \mathbf{A}) \tag{A.52}$$

$$\nabla \cdot (f\mathbf{A}) = f \nabla \cdot \mathbf{A} + (\mathbf{A} \cdot \nabla) f \tag{A.53}$$

$$\nabla \cdot (\mathbf{A} \times \mathbf{B}) = \mathbf{B} \cdot (\nabla \times \mathbf{A}) - \mathbf{A} \cdot (\nabla \times \mathbf{B}) \tag{A.54}$$

$$\nabla \times (\mathbf{A} \times \mathbf{B}) = \mathbf{A}(\nabla \cdot \mathbf{B}) - \mathbf{B}(\nabla \cdot \mathbf{A}) + (\mathbf{B} \cdot \nabla)\mathbf{A} - (\mathbf{A} \cdot \nabla)\mathbf{B} \tag{A.55}$$

$$\nabla \times (f\mathbf{A}) = \nabla f \times \mathbf{A} + f \nabla \times \mathbf{A} \tag{A.56}$$

$$(\nabla \times \mathbf{A}) \times \mathbf{A} = (\mathbf{A} \cdot \nabla)\mathbf{A} - (1/2)\nabla(\mathbf{A} \cdot \mathbf{A}) \tag{A.57}$$

$$\nabla \times (\nabla \times \mathbf{A}) = \nabla(\nabla \cdot \mathbf{A}) - \nabla^2 \mathbf{A} \tag{A.58}$$

Appendix 8: Integral Theorems

Line Integral of a Gradient

$$\int_a^b \nabla f \cdot \mathbf{dl} = f(b) - f(a) \tag{A.59}$$

Divergence Theorem

$$\int_V \nabla \cdot \mathbf{A}\, dV = \oint_S \mathbf{A} \cdot \mathbf{dS} \tag{A.60}$$

Corollaries

$$\int_V \nabla f\ dV = \oint_S f\ \mathbf{dS} \tag{A.61}$$

$$\int_V \nabla \times \mathbf{A}\ dV = -\oint_S \mathbf{A} \times \mathbf{dS} \tag{A.62}$$

Stokes' Theorem

$$\oint_L \mathbf{A} \cdot \mathbf{dl} = \int_S (\nabla \times \mathbf{A}) \cdot \mathbf{dS} \tag{A.63}$$

Corollary

$$\oint_L f\ \mathbf{dl} = -\int_S \nabla f \times \mathbf{dS} \tag{A.64}$$

Tables A to F

Table A Conversion factors for quantities having following dimensions

The dimensions of MLt^{-2} (Force)

Given a quantity in these units ⇩	Multiply by table value to convert to these units ⇨	g cm s^{-2} (dynes)	N = kg m s^{-2} (Newtons)
g cm s^{-2}		1	10^{-5}
N = kg m s^{-2}		10^5	1

The dimensions of $ML^{-1}\ t^{-2}$ (Pressure, Momentum flux)

Given a quantity in these units ⇩	Multiply by table value to convert to these units ⇨	g cm^{-1} s^{-2} (dynes cm^{-2})	kg m^{-1} s^{-2} (N m^{-2}) (Pa)	atm	mm Hg
g cm^{-1} s^{-2}		1	10^{-1}	9.869×10^{-7}	7.501×10^{-4}
kg m^{-1} s^{-2}		10	1	9.869×10^{-6}	7.501×10^{-3}
atm		1.013×10^6	1.013×10^5	1	760
mm Hg		1.333×10^3	1.333×10^2	1.316×10^{-3}	1

(continued)

Table A (continued)

The dimensions of ML^2t^{-2} (Work, Energy, Torque)

Given a quantity in these units ⇓ Multiply by table value to convert to these units ⇨	$g\ cm^2\ s^{-2}$ (ergs)	$kg\ m^2\ s^{-2}$ (absolute joules)	cal	hp-hr	kw-hr
$g\ cm^2\ s^{-2}$	1	10^{-7}	2.390×10^{-8}	3.725×10^{-14}	2.778×10^{-14}
$kg\ m^2\ s^{-2}$	10^7	1	2.390×10^{-1}	3.725×10^{-7}	2.778×10^{-7}
Thermochemical calories	4.184×10^7	4.184	1	1.559×10^{-6}	1.162×10^{-6}
Horsepower hours	2.685×10^{13}	2.685×10^6	6.416×10^5	1	7.457×10^{-1}
Absolute kilowatt-hours	3.600×10^{13}	3.600×10^6	8.604×10^5	1.341	1

The dimensions of $ML^{-1}\ t^{-1}$ (Viscosity, Density times diffusivity, Concentration times diffusivity)

Given a quantity in these units ⇓ Multiply by table value to convert to these units ⇨	$g\ cm^{-1}\ s^{-1}$ (poises)	$kg\ m^{-1}\ s^{-1}$	Centipoises
$g\ cm^{-1}\ s^{-1}$ = (poises)	1	10^{-1}	10^2
$Pa\ s = kg\ m^{-1}\ s^{-1}$	10	1	10^3
Centipoises	10^{-2}	10^{-3}	1

The dimensions of $MLt^{-3}\ T^{-1}$ (Thermal conductivity)

Given a quantity in these units ⇓ Multiply by table value to convert to these units ⇨	$g\ cm\ s^{-3}\ K^{-1}$ ($ergs\ s^{-1}\ cm^{-1}\ K^{-1}$)	$kg\ m\ s^{-3}\ K^{-1}$ ($W\ m^{-1}\ K^{-1}$)	$cal\ s^{-1}\ cm^{-1}\ K^{-1}$
$g\ cm\ s^{-3}\ K^{-1}$	1	10^{-5}	2.390×10^{-8}
$kg\ m\ s^{-3}\ K^{-1}$	10^5	1	2.390×10^{-3}
$cal\ s^{-1}\ cm^{-1}\ K^{-1}$	4.184×10^7	4.184×10^2	1

The dimensions of L^2t^{-1} (Momentum, Thermal, Molecular and Magnetic diffusivities)

Given a quantity in these units ⇓ Multiply by table value to convert to these units ⇨	$cm^2\ sec^{-1}$	$m^2\ sec^{-1}$	Centistokes
$cm^2\ sec^{-1}$	1	10^{-4}	10^2
$m^2\ sec^{-1}$	10^4	1	10^6
Centistokes	10^{-2}	10^{-6}	1

(continued)

Table A (continued)

The dimensions of $Mt^{-3}\ T^{-1}$ (Heat transfer coefficients)

Given a quantity in these units ⇩	Multiply by table value to convert to these units ⇨	$g\ s^{-3}\ K^{-1}$	$kg\ s^{-3}\ K^{-1}$ ($W\ m^{-2}\ K^{-1}$)	$cal\ cm^{-2}\ s^{-1}\ K^{-1}$	$W\ cm^{-2}\ K^{-1}$
$g\ s^{-3}\ K^{-1}$		1	10^{-3}	2.390×10^{-8}	10^{-7}
$kg\ s^{-3}\ K^{-1}$		10^{3}	1	2.390×10^{-5}	10^{-4}
$cal\ cm^{-2}\ s^{-1}\ K^{-1}$		4.184×10^{7}	4.184×10^{4}	1	4.184
$W\ cm^{-2}\ K^{-1}$		10^{7}	10^{4}	2.390×10^{-1}	1

The dimensions of $ML^{-2}\ t^{-1}$ (Mass transfer coefficients)

Given a quantity in these units ⇩	Multiply by table value to convert to these units ⇨	$g\ cm^{-2}\ s^{-1}$	$kg\ m^{-2}\ s^{-1}$
$g\ cm^{-2}\ s^{-1}$		1	10
$kg\ m^{-2}\ s^{-1}$		10^{-1}	1

Table B Constants

1. Mathematical constants	
	$e = 2.71828\ldots..$
	$ln\ 10 = 2.30259\ldots..$
	$\pi = 3.14159\ldots..$
2. Physical constants	
Gas law constant	$R = 1.987$ cal g-mol^{-1} K^{-1}
	$= 82.06$ cm^3 atm g-mol^{-1} K^{-1}
	$= 8.315 \times 10^7$ g cm^2 s^{-2} g-mol^{-1} K^{-1}
	$= 8.315 \times 10^3$ kg m^2 s^{-2} kg-mol^{-1} K^{-1}
Standard acceleration of gravity	$g = 980.665$ cm s^{-2} $= 9.80665$ m s^{-2}
Joule's constant (mechanical equivalent of heat)	$J_c = 4.1840 \times 10^7$ erg cal^{-1} $= 4.1840$ J cal^{-1}
Avogadro's number	$\tilde{N} = 6.02 \times 10^{23}$ molecules g-mol^{-1}
Boltzmann's constant	$k = R/\tilde{N} = 1.380 \times 10^{-16}$ erg K^{-1} $= 1.380 \times 10^{-23}$ J K^{-1}
Stefan-Boltzmann constant	$\sigma = 1.355 \times 10^{-12}$ cal s^{-1} cm^{-2} K^{-4} $=5.671 \times 10^{-8}$ W m^{-2} K^{-4}
Electron charge	$e = 1.602 \times 10^{-19}$ C (coulomb)
Speed of light in a vacuum	$c = 2.99792 \times 10^8$ m sec^{-1}
Permittivity in vacuum	$\varepsilon_0 = 8.85 \times 10^{-12}$ farad m^{-1} (A^2 s^4 kg^{-1} m^{-3})
Permeability in vacuum	$\mu_{m0} = 4\pi \times 10^{-7}$ henry m^{-1} (kg m A^{-2} s^{-2})
Mass of electron	$m_e = 9.108 \times 10^{-31}$ kg

Table C Conversion table for Gaussian MKS units.

The following table gives the relations between the units of common quantities in the rationalized MKS system and in the Gaussian (CGS) system

Quantity	Rationalized MKS	Gaussian
Length	1 meter (m)	$= 10^2$ centimeter (cm)
Mass	1 kilogram (kg)	$= 10^3$ gram (g)
Time	1 second (s)	$= 1$ second (s)
Force	1 Newton (N)	$= 10^5$ dyne
Work (Energy)	1 joule (J)	$= 10^7$ erg
Power	1 watt (W) $= 1\ \mathrm{J\ s^{-1}} = 1\ \mathrm{V \times A}$	$= 10^7$ erg $\mathrm{sec^{-1}}$
Charge	1 coulomb (C)	$= 10^{-1}$ CGSemu $= \widehat{3} \times 10^9$ CGSesu
		$= \widehat{3} \times 10^9$ statcoulomb
Current	1 ampere (A) $= 1\ \mathrm{C\ s^{-1}}$	$= 10^{-1}$ CGSemu $= \widehat{3} \times 10^9$ CGSesu
		$= \widehat{3} \times 10^9$ statampere
Current density	$1\ \mathrm{A\ m^{-2}}$	$= \widehat{3} \times 10^5$ statamp
Electric field	$1\ \mathrm{V\ m^{-1}}$	$= (1/\widehat{3}) \times 10^{-4}$ statvolt
Potential	$1\ \mathrm{V} = 1\ \mathrm{J\ C^{-1}} = 1\ \mathrm{W\ A^{-1}}$	$= 10^8$ CGSemu $= (1/\widehat{3}) \times 10^{-2}$ CGSesu
		$= (1/\widehat{3}) \times 10^{-2}$ statvolt
Capacitance	1 farad (F) $= 1\ \mathrm{CV^{-1}}$	$= 10^{-9}$ CGSemu $= \widehat{3}^2 \times 10^{11}$ CGSesu
		$= \widehat{3}^2 \times 10^{11}$ cm
Resistance	1 ohm (Ω) $= 1\ \mathrm{VA^{-1}}$	$= 10^9$ CGSemu $= (1/\widehat{3}^2) \times 10^{-11}$ CGSesu
		$= (1/\widehat{3}^2) \times 10^{-11}\ \mathrm{s\ cm^{-1}}$
Inductance	$1\ \mathrm{H} = 1\ \mathrm{V\ s\ A^{-1}}$	$= 10^9$ CGSemu $= (1/\widehat{3}^2) \times 10^{-11}$ CGSesu

where $\widehat{3} = 2.9979250 \pm 0.0000010$

Table D Comparison among MKS, Gaussian, CGS emu and CGS esu systems

Quantity	Sign	MKSA system	Gaussian system	CGS emu	CGS esu
Force	F, f	1 N (newton)	10^5 dyne	10^5 dyne	10^5 dyne
Work (energy)	W	1 J (joule)	10^7 erg	10^7 erg	10^7 erg
Power	P	1 W (watt)	10^7erg s^{-1}	10^7erg s^{-1}	10^7erg s^{-1}
Electric potential	V	1 V (volt)	$(1/\widehat{3}) \times 10^{-2}$	10^8	$(1/\widehat{3}) \times 10^{-2}$
Electric field	E	1 V m^{-1}	$10^{-4}/\widehat{3}$	10^6	$10^{-4}/\widehat{3}$
Current	I	1 A (amp)	$\widehat{3} \times 10^9$	10^{-1}	$\widehat{3} \times 10^9$
Magnetic potential	V_m	1 A T (amp-turn)	$4\pi/10$	$4\pi/10 \text{ Gb}$	$4\pi \times \widehat{3} \times 10^9$
Magnetic field intensity	H	1 A T m^{-1}	$4\pi \times 10^{-3}$	$4\pi \times 10^{-3}$ Oe	$4\pi \times \widehat{3} \times 10^7$
Magnetic flux	Φ_m	1 Wb (weber)	$(1/\widehat{3}) \times 10^{-2}$	10^8 Mx	$(1/\widehat{3}) \times 10^{-2}$
Magnetic flux density	B	$1 \text{ Wb m}^{-2} = 1$ T (tesla)	10^4	10^4 G	$(1/\widehat{3}) \times 10^{-6}$
Charge	Q	1 C (coul.)	$\widehat{3} \times 10^9$	10^{-1}	$\widehat{3} \times 10^9$
Charge density	ρ	1 C m^{-2}	$4\pi \times \widehat{3} \times 10^5$	$4\pi \times 10^{-5}$	$4\pi \times \widehat{3} \times 10^{-5}$
Electric resistance	R	1 Ω(ohm)	$(1/\widehat{3}^2) \times 10^{-11}$	10^9	$(1/\widehat{3}^2) \times 10^{-11}$
Inductance	L	1 H (henry)	10^9	10^9	$(1/\widehat{3}^2) \times 10^{-11}$
Capacitance	C	1 F (farad)	$\widehat{3}^2 \times 10^{11}$	10^{-9}	$\widehat{3}^2 \times 10^{11}$
Permittivity of free space	ε_0	$1/(\widehat{3}^2 \times 4\pi) \times 10^{-9} \text{ F m}^{-1}$	1	$(1/\widehat{3}^2) \times 10^{-20} \text{s}^2/\text{cm}^2$	1
Permeability of free space	μ_{m0}	$4\pi \times 10^{-7} \text{ H m}^{-1}$	1	1	$(1/\widehat{3}^2) \times 10^{-20} \text{s}^2/\text{cm}^2$

where $\widehat{3} = 2.9979250 \pm 0.0000010$

Table E Properties of conducting fluid

Liquid metals	Viscosity μ (10^{-3} N·s/m^2)	Density ρ (10^3 kg/m^3)	Magnetic diffusivity ν_m (m^2/s)	Electric conductivity σ (10^6 mho/m)	Thermal conductivity λ (W/m K)	Temp. K
Liquid steel	5.5at 1,500 C	7.08 at 1,600 C	1.09 at 1,600 C	0.730 at 1,560 C	33.4 at 1,600 C	
	4.4 at 1,650 C	6.86 at 1,650 C	1.12 at 1,600 C	0.714 at 1,600 C		
Liquid mercury	1.55	13.55	0.744	1.07	8.8	293
Liquid sodium	0.705	0.927	0.0765	10.4	85	373
NaK Alloy1 (Na 22% K 78%)	0.5	0.85	0.30	2.66	24.3	293
NaK Alloy2 (Na 56% K 44%)	0.6	0.89	0.266	3	25.8	293
Wood's metal (Pb 50%Bi 25% Sn 12.5% Cd 12.5%)	2	1.05	8.85	9	13.4	373
Air (sea level)	1.825×10^{-2}	1	–	–	0.0238	293
Sea water	1	1.025	2×10^5	4×10^{-6}	0.6	293
Slag (Al_2O_3 20%	0.25 at 1,600 C					
CaO 50%	0.14 at 1,700 C					
SiO_2 30%)						

From W.F.Hughes, F.J.Young, in *The Electromagnetodynamics of Fluids* (Wiley, New York, 1966) and in *Kaitei 3 edition kin-zoku deta bukku* (JIM, Maruzen, Tokyo, 1993), p. 16

Table F Properties of liquid metals (MKS units)

Liquid metals (at.no.)	Melting point (C)	Boiling Point (C)	Density ρ (10^3 kg/m^3)	Electrical conductivity σ (10^6 mho/m)	Viscosity μ (10^{-3} N·s/m^2)	Surface tension σ_f (N/m)
Al (13)	660.2	2,950	2.38 at 660 C	5.1 at 657 C	2.9 at 700 C	0.520 at 750 C
			2.26 at 1,100 C	4.31 at 870 C	1.4 at 800 C	
Bi (83)	271.0	1,477	10.0 at 300 C	0.775 at 300 C	1.66 at 304 C	0.376 at 300 C
			9.20 at 962 C	0.651 at 750 C	0.996 at 600 C	0.363 at 500 C
Cd (48)	321	765	8.01 at 330 C	2.97 at 330 C	2.37 at 350 C	0.564 at 330 C
			7.72 at 600 C	2.80 at 600 C	1.54 at 600 C	0.600 at 500 C
Cs (55)	28.5	705	1.84 at 28 C	2.73 at 30 C	0.630 at 43.4 C	–
					0.343 at 210.9 C	
Fe (26)	1,535	2,750	7.08 at 1,600 C	0.73 at 1,560 C	4.4 at 1,650 C	1.85 at 1,550 C
Ga (31)	29.92	1,983	6.093 at 32.4 C	3.86 at 29.8 C	1.894 at 52.9 C	0.735 at 30 C
			5.445 at 1,100 C	3.66 at 46.1 C	0.652 at 806 C	
In (49)	156.4	2,087	7.026 at 164 C	3.43 at 154 C	–	0.340 at 170 C
Pb (82)	327.4	1,737	10.51 at 400 C	1.084 at 327 C	2.116 at 441 C	0.442 at 350 C
			9.81 at 1,000 C	0.816 at1,000 C	1.185 at 844 C	0.431 at 500 C
Li (3)	179	2,403	0.507 at 200 C	2.21 at 230 C	0.592 at 183.4 C	–
			0.441 at1,000 C		0.455 at 285.5 C	
Hg (80)	–38.37	357	13.55 at 20 C	1.016 at 50 C	1.55 at 20 C	0.465 at 20 C
			12.88 at 300 C	0.737 at 350 C	1.01 at 200 C	0.394 at 354 C
K (19)	63.7	760	0.819 at 100 C	7.6 at 64 C	0.515 at 69.6 C	0.086 at 100 C
			0.676 at 700 C	3.18 at 350 C	0.136 at 700 C	

(continued)

Table F (continued)

Rb (37)	39.0	688	1.475 at 39.0 C	4.32 at 50 C 3.64 at 100 C	0.673 at 38 C 0.323 at 220.1 C	–
Ag (47)	960.5	2,212	9.3 at 960.5 C 9.0 at 1,300 C	6.04 at 1,000 C 4.76 at 1,340 C	2.98 at 1,200 C	0.923 at 995 C
Na (11)	97.8	1,621	0.928 at 100 C 0.780 at 700 C	10.35 at 100 C 5.41 at 350 C	0.686 at 103.7 C 0.182 at 700 C	0.206 at 100 C 0.200 at 250 C
Sn (50)	231.9	2,270	6.834 at 409 C 6.64 at 704 C	2.1 at 231.9 C 1.45 at 1,000 C	1.91 at 240 C 1.05 at 600 C	0.526 at 300 C 0.510 at 500 C
Zn (30)	419.5	1,663	6.92 at 419.5 C 6.57 at 800 C	2.83 at 419.5 C 2.80 at 800 C	3.17 at 450 C 1.88 at 700 C	0.785 at 510 C 0.765 at 640 C
NaK Alloy 1 (Na 22% K 78%)	19	1,518	0.886 at 100 C 0.742 at 700 C	3.03 at 50 C 2.44 at 200 C	0.546 at 103.7 C 0.161 at 700 C	0.100 at 19 C 0.110 at 250 C
NaK Alloy 2 (Na 56% K 44%)	–11	784	0.847 at 100 C 0.703 at 700 C	2.66 at 50 C 2.13 at 200 C	0.468 at 103.7 C 0.146 at 700 C	0.110 at –11 C 0.120 at 250 C
Wood's metal	70	–	9.8 at 73 C 6.94 at 687 C	0.993 at 85.6 C 0.936 at 327 C	281 at 100 C 177 at 200 C	0.326 at 100 C
NaCl standard solution	–	–	1.201 at 18 C	0.226 at 20 C 0.277 at 30 C	108 at 18 C 112 at 60 C	0.08 at 20 C

From W.F.Hughes, F.J.Young, in *The Electromagnetodynamics of Fluids* (Wiley, New York, 1966) and in *Kaitei 3 edition kin-zoku deta bukku* (JIM, Maruzen, Tokyo, 1993), p. 16

Notations

(Number 1 to 5 show the chapters where the notations are used)

A	surface area (m^2) 2
A	magnetic gravity ratio number (-) 5
A'	relative magnetic gravity ratio number (-) 5
A, A^*	complex number (-) 3
$\boldsymbol{A}$	vector potential of a magnetic flux (T m) 4
a	radius of curvature of a metal drop (m) 4
a_1, a_2	real number (-) 3
B, B_e	magnetic flux density (T) 1, 2, 3, 4, 5
B, B^*	complex number (-) 3
$\boldsymbol{B}$	magnetic flux density vector (T) 3, 4, 5
Bi	Bingham number (-) 2
B_{max}, B_{min}	maximum and minimum amplitude of magnetic flux density (T) 4
B_x, B_y, B_z	magnetic flux density components (T) 3, 4
B_0	magnetic flux density (T) 4
Bo	Biot number (-) 2
b_1, b_2	real number (-) 3
b_z, b_x, $\widehat{b}_z$, $\widehat{b}_x$, b_0	magnetic flux density (T) 3, 4
b_x^*, b_z^*	complex conjugate numbers of b_x and b_z, respectively 3
ber_ν, bei_ν	Kelvin function of order ν(-) 3
C	circulation (m^2/s) 3
c	concentration (kg/m^3) 2
c_1	constant $\{kg/(m^2\ s^2)\}$ 3
c_b	bulk concentration (kg/m^3) 2
c_s	surface concentration (kg/m^3) 2
c_p	heat capacity at constant pressure $\{J/(kg\ K)\}$ 1, 2, 3
D	diffusion coefficient (m^2/s) 2
$\boldsymbol{D}$	displacement field (C/m^2) 1, 2, 3

S. Asai, *Electromagnetic Processing of Materials*, Fluid Mechanics and Its Applications 99, DOI 10.1007/978-94-007-2645-1,
© Springer Science+Business Media B.V. 2012

D_I	Damköhler-I number (-) 2
D_{II}	Damköhler-II number (-) 2
D_{III}	Damköhler-III number (-) 2
E	energy density (J/m^3) 5
$\boldsymbol{E}$	electric field (V/m) 1, 2, 3
$E_x,\ E_y,\ E_z$	electric field components (V/m) 3
$\boldsymbol{F}$	body force vector (N/m^3) 4, 5
F	body force (N/m^3) 2
F	function showing the shape of boundary (-) 3
Fo	Fourier number (-) 2
Fr	Froude number (-) 2
Fr_m	magnetic Froude number (-) 5
$f, \boldsymbol{f}$	external body force (N/m^3) 2, 4
f_{int}	intermittent frequency (1/ s) 4
$\bar{f}_z,\ \bar{f}_r$	time average body force (N/m^3) 3
f_x, f_y, f_z	body force components (N/m^3) 3, 4
f	friction coefficient (-) 2
Gr	Grashof number (-) 2, 3
Gr_m	magnetic Grashof number (-) 5
$g, \boldsymbol{g}$	gravitational acceleration coefficient (m/s^2) 2, 4, 5
$\boldsymbol{g}'$	apparent gravitational acceleration coefficient (m/s^2) 4
H	magnetic field intensity (A/m) 1, 2, 5
$\boldsymbol{H}$	magnetic field intensity vector (A/m) 3, 5
Ha	Hartmann number (-) 2, 3
H_d	demagnetizing magnetic field intensity (A/m) 5
H_{eff}	magnetic field intensity induced in a substance (A/m) 5
H_{ex}	magnetic field intensity imposed from the outside of substance (A/m) 5
H_s	magnetic field intensity at which a magnetic moment is saturated (A/m) 5
h	heat transfer coefficient {W/(m^2 K)} 2
h	surface wave height (m) 2
h	surface height (m) 4, 5
h_0	liquid depth (m) 2
I_z, I_{sc}	current (A) 3
i	imaginary number (-) 3, 4
$\boldsymbol{i_x}, \boldsymbol{i_y}, \boldsymbol{i_z}$	unit vector in *x*-, *y*- and *z*-directions, respectively (-) 3
J	electric current density (A/m^2) 3
$\boldsymbol{J}$	mass flux {kg/(m^{2}s)} 2
$\boldsymbol{J}$	electric current density (A/m^2) 3, 4
J_s	surface current (A/m) 3
$J_x,\ J_y,\ J_z,\ J_\theta$	electric current density (A/m^2) 3, 4
J_x	mass flux in the *x*-direction {kg/(m^2 s)} 2
J_p	Bessel function of the first kind of order p (-) 3

j_y	electric current density (A/m^2) 3
$j_y{}^*$	conjugate of j_y 3
k	average mass transfer coefficient (m/s) 2
k	constant (1/ m) 3
k	Boltzmann constant (1.38×10^{-23} J/K) 5
k'	instantaneous mass transfer coefficient (m/s) 2
L, L'	characteristic length (m) 2, 3
L	falling length of particle (m) 5
Le	Lewis number (-) 2
L_m	characteristic length relating with magnetic gradient (m) 5
l	diffusion length of mass (m) 2
$\boldsymbol{M}$	magnetic moment (A/m) 5
M_m	magnetic moment of ferromagnetic materials (A/m) 5
M_n	magnetic moment in non-magnetic materials (A/m) 5
Ma_m	magnetic Mach number or Alfvén number (-) 3
M_s	saturated magnetic moment (A/m) 5
m_{mol}	molecule mass (g/mol) 5
N	Stuart number or interaction parameter (-) 2, 3
N	demagnetizing factor (-) 5
N_x, N_y, N_z	components of demagnetizing factor in x-, y- and z-directions, respectively (-) 5
Nu	Nusselt number (-) 2
n	constant (1/ m) 3
$\boldsymbol{n}$	unit normal vector (1/ m) 3
P	pressure (Pa) 2, 3
Pe	heat Peclet number (-) 2
Pe^*	*m*ass Peclet number (-) 2
P_m	magnetic pressure number (-) 3
Pr	Prandtl number (-) 2
Pr_m	magnetic Prandtl number (-) 3
p_c	pressure due to surface tension (Pa) 4
p_{dyn}	dynamic pressure due to molten metal motion (Pa) 4
p_s	static pressure of molten metal (Pa) 4
p_m	magnetic pressure (Pa) 3, 4
P	pressure (Pa) 3
Q	heat generation per unit volume (W/m^3) 2
Q	volumetric flow rate (m^3/s) 2
Q	Ohmic dissipation per unit area (W/m^2) 3
$\boldsymbol{q}$,q	heat flux { J/(m^2 s)} 2, 4
q_x	heat flux in the x-direction { J/(m^2 s)} 2
$\bar{q}$	Ohmic dissipation (J/m^3) 3, 4
R	Reaction term {kg/(m^3 s)} 2
R	half diameter of a cylindrical tube and spherical ball (m) 2
Ra	Rayleigh number (-) 2

Ra_m	magnetic Rayleigh number (-) 5
Re	Reynolds number (-) 2
Re_m	magnetic Reynolds number (-) 3
R_i	electric inner resistance (Ω) 3
r	axis in circular cylindrical and spherical coordinates (-) 2
r	average curvature radius (m) 3
r	particle size (m) 5
Sc	Schmidt number (-) 2
Sh	Sherwood number (-) 2
T	temperature (K) 2
T	period of a periodic function (s) 2, 4
T	magnetization torque (N m) 5
$\boldsymbol{T}$	stress tensor (N/m^2) 3
T_b	temperature in bulk flow (K) 2
T_s	temperature at interface (K) 2
t	time (s) 2, 3, 5
t_e	contacting time of two fluids (s) 2
t_{rel}	mechanical relaxation time (s) 4
U	magnetization energy (J/m^3) 5
U	electromotive force (V) 4
$U_{a,b}$, U_c	magnetization energy in a- or b-axis and c-axis, respectively (J/m^3) 5
V	volume of a vessel (m^3) 2
V_T, V_{oc}, V_g	voltage (V) 3
v	velocity (m/s) 2
$\boldsymbol{v}$	velocity vector (m/s) 3, 4, 5
v_r, v_θ, v_x, v_y, v_z	velocity components (m/s) 3, 4
$\overline{v}_x$	average flow velocity (m/s) 3
W	energy density (J/m^3) 1, 2
We	Weber number (-) 2
W_m	shielding parameter or magnetic Womersley number (-) 3
w	mass transfer rates (kg/s) 2
x	axis in rectangular (Cartesian) coordinates (-) 2, 3, 4
x	variable (-) 3
Y	Yakob number (-) 2
y	axis in rectangular (Cartesian) coordinates (-) 2, 3, 4
z	axis in rectangular (Cartesian) coordinates (-) 2, 3, 4
β	thermal coefficient of volumetric expansion (1/ K) 2
β	constant ($\beta^2 \equiv \gamma^2 + i\omega/v_m$) (1/m) 3, 4
β_r, β_i	real and imaginary parts of β (1/m) 3, 4
γ	propagation constant (1/m) 3, 4
ΔH	enthalpy change (J/kg) 2
δ	skin depth (m) 3, 4
δ_{ij}	Kronecker delta (-) 3
ε	dielectric constant (F/m) 3

θ	angle in circular cylindrical and spherical coordinates (-) 2
θ	angle between the easy magnetization axis and the imposed magnetic field (rad) 5
κ	thermal diffusivity (m^2/s) 2
μ	viscosity coefficient $\{kg/(m\,s)\}$ 2, 3, 5
μ_m	magnetic permeability (H/m) 1, 2, 3, 4
μ_{m0}	magnetic permeability in vacuum, $4\pi \times 10^{-7}$(H/m) in MKSA unit system, 1(-) in cgs Gauss unit system 5
μ_r	relative magnetic permeability ($\mu_r \equiv 1 + \chi$)(-) 5
ν	kinematic viscosity (momentum diffusivity) (m^2/s) 2, 3, 5
ν_m	magnetic diffusivity (m^2/s) 3, 4
ξ	wave number (1/m) 2
λ	thermal conductivity { J/(m s K)} 2, 3
λ	wave length (m) 2
λ	angular velocity (rad/s) 3
λ_n	n-th root of $J_0(\lambda_n R) = 0$ (1/m) 2
ρ	density (kg/m^3) 1, 2, 3, 4, 5
ρ''	density (g/cm^3) 5
ρ_f	net charge density (C/m^3) 3
ρ_l, ρ_p	densities of liquid and particle, respectively (kg/m^3) 5
ρ_e, ρ_{es}	electric charge (C) 3
σ	electric conductivity (S/m) 2, 3, 4
σ_f	surface tension (N/m) 2, 3, 4
σ_{ij}	stress tensor $\{kg/(m\,s^2)\}$ 2, 3
$\boldsymbol{\tau}$	momentum flux $\{kgm/s\}/(m^2\,s)$ 2
τ_0	stress (N/m^2) 2
τ_R	crystal alignment time in synchronized region (s) 5
τ_s	crystal alignment time (s) 5
τ_{yx}	flux of the x-momentum in the negative y-direction $\{kg/(m\,s^2)\}$ 2
Φ	velocity potential (m^2/s) 2
ϕ	magnetic scalar potential (T m) 3
φ	angle in spherical coordinates (-) 2
χ	magnetic susceptibility (-) 1, 2, 5
$\boldsymbol{\chi}$	magnetic susceptibility tensor (-) 5
$\chi_{a,b}$, χ_c	magnetic susceptibilities in a- or b-axis and c-axis, respectively (-) 5
χ_r	relative susceptibility (-) 5
χ'	magnetic susceptibility in SI for E-H unit system (H/m) 5
χ''	magnetic susceptibility in cgs Gauss unit system (-) 5
$\chi_\perp$	magnetic susceptibility in the difficult magnetization axis (-) 5
χ_g	mass magnetic susceptibility (m^3/kg) 5
χ''_g	mass magnetic susceptibility ($c\,m^3/g$) 5
χ_m	molar magnetic susceptibility ($m^3/k\,mol$) 5

χ''_m	molar magnetic susceptibility (c m^3/mol) 5
χ_{med}	magnetic susceptibility of medium (-) 5
χ_p	magnetic susceptibility of precipitated substance (-) 5
ω	angular frequency (rad/s) 2, 3, 4
ω	rotation speed of magnetic field (rad/s) 5
$\boldsymbol{\omega}$	vorticity vector (1/s) 3, 5
ω_r, ω_θ, ω_z	vorticity components (1/s) 3

Index

S. Asai, *Electromagnetic Processing of Materials*, Fluid Mechanics and Its Applications 99, DOI 10.1007/978-94-007-2645-1,
© Springer Science+Business Media B.V. 2012

The manufacturer's authorised representative in the EU is Springer Nature Customer Service Centre GmbH, Europaplatz 3, 69115 Heidelberg, Germany. If you have any concerns regarding our products, please contact ProductSafety@springernature.com

Printed and bound by CPI Group (UK) Ltd, Croydon, CR0 4YY
24/07/2026
02175900-0013